Earth Science

49 Science Fair Projects

Other Books in the
Science Fair Projects Series

BOTANY:

49 Science Fair Projects (No. 3287)

This first volume in the series concentrates on plant germination, photosynthesis, hydroponics, plant tropism, plant cells, seedless plants, and plant dispersal.

ENVIRONMENTAL SCIENCE:

49 Science Fair Projects (No. 3369)

This third volume in the series deals with Earth's surroundings and how pollution, waste disposal, irrigation, errosion, and heat and light affect the ecology.

Earth Science

49 Science Fair Projects

Robert L. Bonnet
and
G. Daniel Keen

This book is dedicated to our loving children, Margie and Bob Bonnet, and Alicia and Trisha Keen. We love you.

FIRST EDITION
FIRST PRINTING

Library of Congress Cataloging-in-Publication Data

Bonnet, Robert L.
Earth science : 49 science fair projects / by Robert L. Bonnet and G. Daniel Keen.
p. cm.
ISBN 0-8306-9287-8 ISBN 0-8306-3287-5 (pbk.)
1. Earth sciences—Experiments. 2. Earth sciences—Exhibitions. 3. Science projects. I. Keen, G. Daniel. II. Title.
QE44.B66 1989 89-39634
CIP

TAB BOOKS Inc. offers software for sale. For information and a catalog, please contact TAB Software Department, Blue Ridge Summit, PA 17294-0850.

Questions regarding the content of this book should be addressed to:

Reader Inquiry Branch
TAB BOOKS Inc.
Blue Ridge Summit, PA 17294-0214

Acquisitions Editor: Kimberly Tabor
Book Editor: Lori Flaherty
Production: Katherine Brown
Illustrations: Carol Chapin

Contents

Acknowledgments

We wish to extend our appreciation to Bob Blough, Mark Chamberl, and W. Daniel Keen, R.P.

Introduction

At some time during our school years, all of us have been required to do at least one science project. It might have been growing seeds in kindergarten or building a Tesla coil in high school, but such experiences are not forgotten and help shape our views of the world around us.

Doing a science project yields many benefits beyond the obvious educational value. The logical process required helps encourage clear, concise thinking, which can carry through the student's entire life. Science in general requires a discipline of the mind, clear notes and data gathering, a curiosity and patience, an honesty regarding results and procedures, and last, concise reporting of work accomplished. A student's success in a science project can provide him or her with the motivation to strive for success in other areas.

Often parents are encouraged to work with their children on projects, thus fostering richer family relationships as well as enhancing the child's self-esteem.

Finally, there is always the possibility of a spin-off interest developing. For instance, a student might choose to do a project in mathematics, perhaps using a battery, switches, and light bulbs to represent binary numbers and discover a liking for electronics.

But where does the student, the teacher, or the parent look for suggestions for such projects? It is our aim to address this void by offering a large collection of projects and project ideas in a series of books. These books target anyone who wants or needs to do a science project. A teacher might want to do classroom projects; a student might be assigned to do a project for class by a science teacher, or to enter one in a science fair; a parent might want to help their child with a science fair project; or someone might want to do an experiment or project just for the fun of it. Science teachers can use these books to help them conduct a science fair at their school or to suggest criteria for judging. Parents might feel apprehensive when their child comes home with a project requirement to do for school. These books will come to their rescue.

Our goal is to provide you with project ideas from beginning to end. Students need a starting place and direction. The questions in this book are posed in the form of needs or problems (discovering how to get electricity from the sun was born out of a need, for example). Overviews, organizational direction, suggested hypotheses, materials, procedures, and controls are provided. The projects explained are complete but can also be used as spring boards to create expanded projects. All projects are brainstormed for going further. The reader will be shown how to develop ideas and projects using valid scientific processes and procedures.

The chapters are organized by topics. Some projects might overlap into more than one science discipline, as well as within the discipline. The attempt has been made to place such projects under the seemingly dominant theme. You can quickly skim through the topics and "home in" on a project suitable for your ability group and interests.

Projects are designed for the sixth- to ninth-grade student. Many projects can be "watered down," however, and used for children in grades lower than sixth grade. Similarly, students in upper high school grades can take each project's brainstormed ideas to more advanced levels.

It is very important to read the introduction given at the beginning of the chapter from which you plan on doing a project. Information that is relevant to each project in the chapter is given here. The Appendix contains a list of suppliers where you can purchase laboratory supplies.

After you have selected a project and read the chapter's introduction, read the entire project through carefully. This will help you understand the overall scope of the project, the materials needed, the time requirements, and the procedure before you begin.

Safety and ethics must always be a consideration. Some projects require cutting with a knife or scissors, and common sense usage should be practiced. Projects that *must* have adult supervision are indicated with the phrase **Adult supervision required** next to the title. These projects deal with caustics, poisons, acids, high temperatures, high voltages, or other potentially hazardous conditions. Ethical science concepts involve very careful considerations about living organisms. One should not recklessly cause pain, damage, or death to any living organism.

There is no limit to the number of themes and the number of hypotheses one can form about our universe. The number is as infinite as the stars in the heavens. It is our hope that many students will key off some of the ideas presented, develop their own unique hypotheses, and proceed with their experiments using accepted scientific methods.

Some projects could go on for years. There is no reason to stop a project, other than getting tired of it. It might be that what you studied this year can be taken a step further next year. With each question or curiosity answered, more questions are raised. It has been our experience that answers produce new and exciting questions. We believe that science discovery and advancement proceeds as much on excellent questions as it does on excellent answers.

We hope we can stimulate the imagination and encourage creative thinking in students, teachers, parents, and the public at large. Learning is rewarding and enjoyable. Good luck with your project!

Robert L. Bonnet
G. Daniel Keen

How to Use This Book

All projects that require adult supervision have the STOP symbol at the beginning of the project. No responsibility is implied or taken for anyone who sustains injuries as a result of the materials or ideas put forward in this book. Taste nothing that is not directly food related. Use proper equipment (gloves, safety glasses, and other safety equipment). Clean up broken glass with a dust pan and brush. Use chemicals with extra care. Wash hands after project work is done. Tie up loose hair and clothing. Follow step-by-step procedures; avoid short cuts. Never work alone. Remember, adult supervision is advised. Use common sense and make safety the first consideration, and you will have a safe, fun, educational, and rewarding project.

1

Science Projects

Before you begin to work on a science project, there are some important things to know. It is important that you read this chapter before starting on a project. It defines terms and sets up guidelines that should be adhered to as work on the project progresses.

Beginning without Pain

A person cannot proceed with a science project unless the term "science project" is fully understood. Older students are familiar with report writing. Many types of reports are required at all grade levels, whether it is book reports, history reports, or perhaps term papers. Although a report might be required to accompany a science project, it is not the focal point. The body of science comes from experimentation. Most projects discover information by scientific methods. The "scientific method" is a formal approach to scientific investigation. It is a step-by-step logical thinking process. The scientific method can be grouped into the following sections:

1. The statement of the problem.
2. The hypothesis.

3. Experimentation and information gathering (results).
4. A conclusion based on the hypothesis.

First, a statement of the problem must be made. A "problem" for scientists does not mean that something went wrong. A problem is something for which there is no good answer. Air pollution is a problem. Aggressive behavior, crab grass, and obesity are problems. Any question can be stated as a problem. Discuss your ideas with someone else, a friend, teacher, parent, or someone working in the field being investigated.

A hypothesis is an educated guess. It is educated because more than likely you know a little about whatever the subject matter might be, such as trees, dogs or bugs. Your life experiences help you form a specific hypothesis rather than a random one. Suppose you hypothesize, "If I add sugar to water and feed it to this plant, it will grow better." You would first need a "control" plant that was given only water. Both plants would be given the identical amount of sunshine, water, temperature, and any other nonexperimental factors.

Assumptions

Assumptions must be defined. What is meant by saying "The plant will grow better?" What is "better" assumed to be? Does it mean greener leaves, faster growing, bigger foliage, better tasting fruit, more kernels per cob?

When growing plants from seeds, the assumption is made that all the seeds are of equal quality. When several plants are used in an experiment, it is assumed that all the plants are the same at the start of the project.

When doing a project, be sure to state all your assumptions. If the results of an experiment are challenged, the challenge should be on the assumptions and not on the procedure.

Sample Size

"Sample size" refers to the number of items in a test. The larger the sample size, the more significant the results.

Using only two plants to test the sugar theory would not yield a lot of confidence in the results. One plant might have grown better than the other because some plants just grow better than others! Obviously, our statistical data becomes more meaningful as we sample a larger group of items in the experiment. As the group size increases, individual differences have less importance.

Making accurate measurements is a must. The experimenter must report the truth and not let bias (his or her feelings) affect his measurements. As we mentioned earlier, the reason science progresses is because we do not have to reinvent the wheel. Science knowledge builds on what people have proven before us. It is important to document (record) the results. They must be replicable (able to be repeated), so that others can duplicate our efforts. Good controls, procedures, and clear record keeping are essential.

As information is gathered, the results could lead to further investigation. More questions might come to light that need asking.

The conclusion must be related to the hypothesis. Was the hypothesis correct or incorrect? Perhaps it was correct in one aspect but not in another. In the sugar example, adding sugar to the water might have helped but only to a point. The human body can use a certain amount of sugar for energy, but too much can lead to health problems.

There is no failure in a science experiment. The hypothesis might be proven wrong, but learning has still taken place. Information has been gained. Many experiments prove to be of seemingly no value, except that someone reading the results will not spend the time to repeat the experiment. This also brings out the point that it is important to thoroughly report results. Mankind's knowledge builds upon success and failure.

Collections, Demonstrations, and Models

Competitive science fairs usually require experimentation. Collections and models by themselves are not experiments, however, they can be turned into experiments. A collection is simply gathered data. Suppose a collection of shells has been assembled from along the eastern seaboard of the United States. The structure and composition of shells from the south can be compared to those found in the north.

The collection then becomes more experimental. Similarly, an insect collection can deal with insect physiology or comparative anatomy. A rock and mineral collection might indicate a greater supply of one type over another because of the geology of the area where they were collected. Leaves could be gathered from trees to survey the available species of trees in your area.

Classroom assignments can be served well by demonstrations. Models can help students to better understand scientific concepts. A steam engine dramatically shows how heat is converted to steam and steam is converted into mechanical energy. Seeing this happen can have a greater educational impact than merely talking about it.

Individuals versus Group Projects

A teacher might require a group of students to work on a project, although they can be difficult for a teacher to evaluate. Who did the most work? If it is dealt with on an interest level, however, then the more help received, the better the project can be. Individual versus group projects bear directly on the intended goal. Most science fairs do not accept group projects.

Choosing a Topic

Select a topic of interest; something that arouses curiosity. One only needs to look through a newspaper to find a contemporary topic: dolphins washing up on the beach, the effect of the ozone layer on plants, stream erosion, etc.

Limitations and Precautions

Of course, safety must always be placed first when doing a project. Using voltages higher than those found in batteries have the potential for electrical shock. Poisons, acids, and caustics must be carefully monitored by an adult. Temperature extremes, both hot and cold, can cause harm. Objects that are sharp or can shatter, such as glass, can be dangerous. Nothing in chemistry should require tasting! Combinations of chemicals can produce toxic materials. Safety goggles, aprons, heat gloves, rubber gloves for caustics and acids, vented hoods, and adult supervision are safety considerations. Each project should be evaluated for the need of these safety materials. Projects in this book that require adult supervision are indicated as such in the project title. Special considerations are important if a project is to be left unattended and accessible to the public.

Most science fairs have ethical rules and guidelines. Live animals, especially vertebrates, are given thoughtful consideration. One might be required to present a note from a veterinarian or other professional that proper handling of the animal has been dealt with. An example would be using mice to run through a maze, demonstrating learning or behavior.

Limitations on time, help, and money are important factors. The question of how much money is to be spent should be addressed by a science fair committee. When entering a project in a science fair, generally the more money spent on the display the better the chance of winning. It isn't fair that one child might only have $1.87 to spend on a project because of family income while another might have $250. Unfortunately, at many science fairs, the packaging does influence the judging. An additional problem might be that one student's parent is available to help while another student's parent might be unavailable to help.

Science Fair Judging

In general, science fairs lack well-defined standards. The criteria for evaluation can vary from school to school, area to area, and region to region. We would like to propose some goals for students and teachers to consider when judging.

A good science project should require creative thinking and investigation by the student. Record keeping, logical sequence, presentation, and originality are important points.

The thoroughness of the student's project reflects the background work that was done. If the student is present, a judge can orally quiz to see if the experimenter has sufficient understanding. Logging all experiences, such as talking to someone knowledgeable in the subject or reading material on it, will show the amount of research put into the project.

Clarity of the problem, assumptions, procedures, observations, and conclusions are important judging criteria too. Be specific.

Points should be given for skill. Technical ability and workmanship are necessary to a good project. Skill categories could include computation, laboratory work, observation, measurement, construction, and other skills.

Often, projects with flashier display boards do better. Some value should be placed on dramatic presentation, but it should not carry the point weight of other criteria, such as originality. Graphs, tables, and other illustrations can be good visual aids to the interpretation of data. Photographs are especially important for projects where it is impossible to set the project up indoors (a "fairy ring" of mushrooms in the forest, for example).

Some science fairs require a short abstract or synopsis in logical sequence. It should include the purpose, assumptions, a hypothesis, materials, procedure, and conclusion.

Competing

Often, your project must compete with others, whether it is at the class level or at a science fair. Find out ahead of time what the rules are for the competition. Check to see if there is a limit on size. Is an accompanying research paper required? Will it have to be orally defended? Will the exhibit have to be left unattended overnight? Leaving a $3,000 computer unattended overnight would be a big risk.

Find out which category has the greatest competition. You might be up against less competition by placing your project in another category. If it is a "crossover" project, you might want to place it in a category that has fewer entries. For example, a project that deals with chloroplasts could be classified under botany or chemistry. A project dealing with the wavelength of light hitting a plant could be botany or physics.

Earth Science

This book deals with a wide range of ideas in the earth science category. Earth science is the scientific study of all nonliving natural-forming materials of the earth. The field of earth science includes a wealth of broad categories such as weather, climate, oceans, rocks, minerals, the earth's surface, geology, archaeology, and energy resources. Knowing how the earth works is important to man's existence on it.

The forces that are at work constantly changing the earth must be studied for man to plan his future and be warned of impending danger.

Complete projects are shown under each earth science category. Many additional ideas for going further are provided for you to develop and investigate. Most of the equipment and supplies required for these projects are inexpensive items or can be found around the home. A Resource List is provided in the back of the book for supplies required for some projects.

Pick a project that is at your level of ability and narrow enough for you to accomplish. It should not require equipment or supplies beyond what can be obtained. For example, a project requiring lodgepole pine cones should not be attempted if these cones are not available.

We hope you get a good feeling of accomplishment as you delve into the fascinating world of earth science.

2

The Earth's Crust

The earth's crust is called the lithosphere. It is the outer-most layer of the solid earth. The temperatures and pressures of the crust are relatively low. The entire crust makes up one percent of the earth's volume but only four-tenths percent of the earth's mass. Therefore, the material of the earth's crust is considerably less dense than the other layers of earth. Because it is less dense, it floats on top of the other layers.

With a diameter of 8,000 miles, the earth's crust is a mere twenty-five miles thick. This thin outer covering, however, varies in thickness from one place to another.

PROJECT 1
Plated Guesses

Overview

Most scientists believe that a long time ago, the seven continents (North America, South America, Antarctica, Europe, Asia, Africa, and Australia) were joined together to make one huge land mass. They theorize that this supercontinent, referred to as Pangaea, broke into pieces and began moving apart. These parts, or "plates," are drifting as the earth's crust drifts on the liquid core underneath. The theory and study of these plates is called plate tectonics.

Do the continents fit comfortably together like the pieces of a jigsaw puzzle to form one large land mass? You can cut out shapes of the seven continents and try to piece them together. Hypothesize that the model can show an accurate account of the plate theory.

Materials

- globe
- construction paper
- tracing paper
- pen or pencil
- scissors
- research materials on Pangaea theory

Procedure

Lay tracing paper over the continents on a globe. Trace the outline of the continents (see Fig. 2-1). Use the traced outlines as templates to cut out continent shapes from the construction paper. Attempt to piece them together to form Pangaea.

Having formed Pangaea, use research material to compare your results to those models constructed by other scientists. Be aware that erosion has taken place over the years. Also, the level of the oceans may have changed so the pieces may fit perfectly. Conclude whether or not your hypothesis was correct or incorrect.

Fig. 2-1. *Trace the continents from the globe on tracing paper. Cut them out and try to form Pangaea.*

PROJECT 2
High in the Sky

Overview

In 1989, a seashore resort town in New Jersey placed a ban on building any more high-rise apartment buildings because of the beach erosion they caused. The tall buildings deflect winds and cause moving air to go around and between them. When a volume of air has to pass through a smaller area, its velocity increases. In traveling further, the wind must travel faster.

When two high-rise buildings are next to each other with only a small gap between them, the wind hitting the buildings directly is sucked through the gap by the air that was already moving through the gap (see Fig. 2-2). This is called the Venturi Effect.

This experiment simulates in the laboratory the effect a high-rise building might have on beach erosion. Hypothesize that moving air will pick up sand particles in front of the building or structure and deposit it to the sides behind the structure.

Materials

- wooden box frame about 2 feet by 4 or 5 feet and 1 or 2 inches deep
- beach sand or playground sand
- fan, preferably one with three speeds
- brick

Procedure

Construct a sandbox out of wood about two feet wide by four or five feet long. It can be shallow, with a depth of an inch or two. Fill the box with fine sand particles from a beach or playground. The sand should be level. Set an electric fan at one end of the box. Stand a brick upright about one foot in from the fan, as shown in Fig. 2-2).

Let the fan blow for a length of time. Observe any sand erosion and any places where sand is being deposited. From your observations, conclude whether your hypothesis was correct.

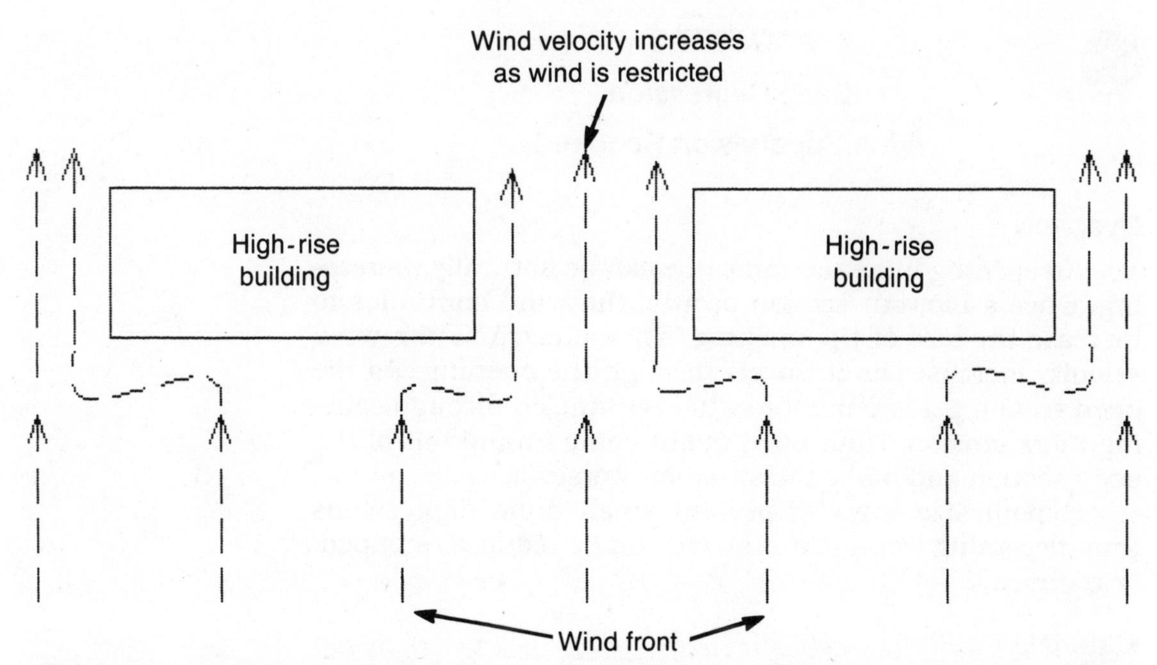

Fig. 2-2. *Air speed increases as it is squeezed through a gap between high-rise buildings. This is called the Venturi effect.*

Going Further

1. Inspect a high-rise building along a beach. Do the corners of the building show any signs of erosion? If so, could you invent some kind of replaceable corners to protect the building?
2. Calculate the different wind velocities at various points around high-rise buildings using a hand-held anemometer.

STOP

PROJECT 3
Deep Depression
Adult Supervision Required

Overview

An opening in a sand dune line can be naturally increasing. Once a blowout section occurs, the wind continues to increase the size of the opening. This is because the wind velocity increases as it travels through the opening. As the open section get larger, more wind is funneled through causing more erosion. High tides might come in and out of the open section and make the situation worse.

Hypothesize ways to prevent small dune depressions from becoming worse. Can the erosion be reduced, stopped, or reversed?

Materials

- wooden box frame (about 2 feet by 4 or 5 feet, and 3 to 4 inches deep)
- beach sand or playground sand
- fan, preferably one with several speeds
- popsicle sticks, model bushes or trees, and other materials that might prevent wind erosion of dunes

Procedure

Construct a wooden box frame about two feet wide by four or five feet long. It should be at least three inches deep. Fill it with sand and level it. In the middle, build a dune line perpendicular to the flow of air. Dig a slight opening in the sand dune. Set up a fan at one end of the box (see Fig. 2-3). Let the fan blow over the box frame and observe the depression. Does it grow bigger?

Try various techniques to prevent the depression from widening. You might use a scale model bush to simulate growing vegetation such as dune grass. Try "planting" bushes on the sides of the depression to trap sand particles. As the sand accumulates along the sides, more vegetation might need to be planted.

What would you do if winter were coming and you could not plant vegetation? Where would the deposition of sand occur on the other side of the blowout?

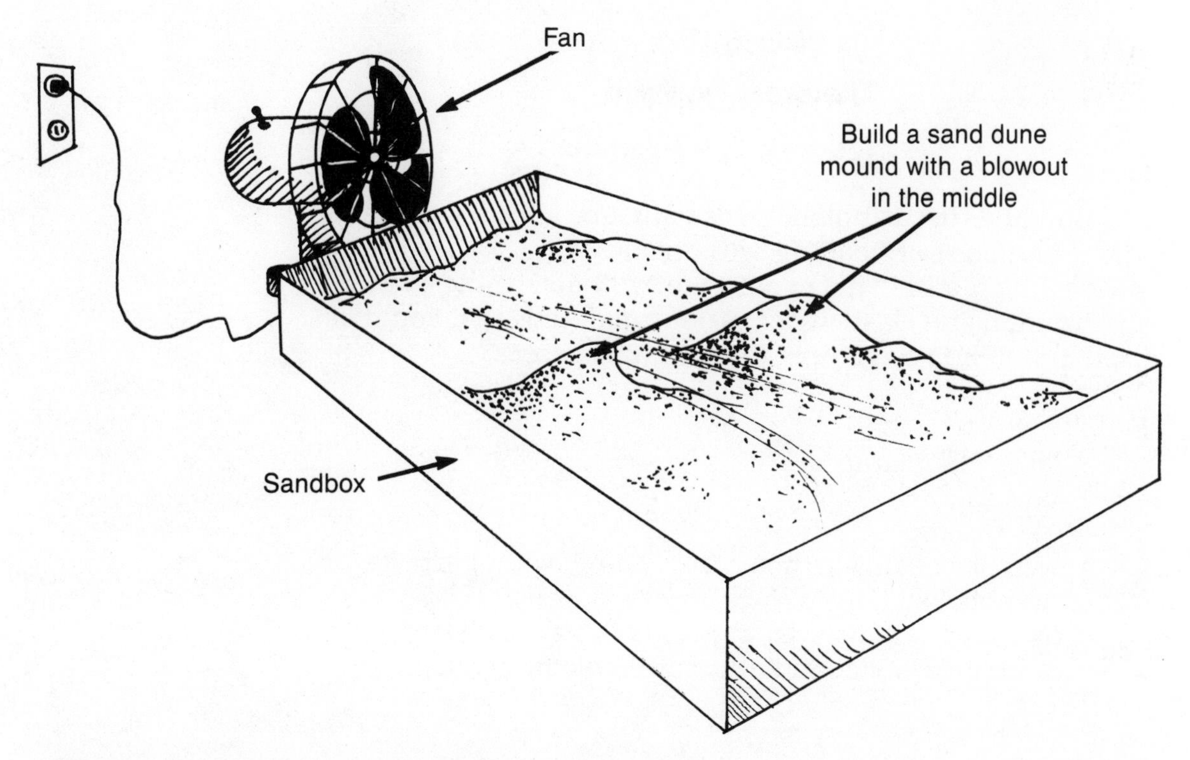

Fig. 2-3. *A depression in a sand dune is self-eroding.*

Going Further

Go to a sand dune at a beach and locate a blowout or opening in a dune line. Thumbtack a one inch wide strip of silk to a sixteen-inch stake. Make six of these. Place two stakes in the blowout. To the right and left of the blowout, place two stakes on each side, a little apart. While the wind is blowing, take a photograph standing at one end to show all the silk pieces. From the photograph, determine where the wind velocity is the greatest.

PROJECT 4

The Breaking Point

Overview

Buildings in earthquake areas must be able to withstand a great amount of bending without breaking if they are to survive the quake. This experiment will determine the tensile strength and elasticity of various materials that could be used in building structures in earthquake areas.

Materials

- large "C" clamp
- tabletop
- piece of rope
- weights such as those that come with barbells or dumbbells
- four foot long boards: balsa wood, particleboard, oak, laminated wood, and other available materials
- ruler

Procedure

Take a four foot long piece of balsa wood, about one inch wide and a quarter-inch thick. Using a "C" clamp, clamp one end of the board to the tabletop and let the rest of the board hang off of the table. Beginning with the lightest weight available, hang weights from the end of it (see Fig. 2-4). By hanging weights near the floor, the weights do not have far to drop when the board breaks. Measure the distance it bends before being broken (bent beyond the point from which it will flex back). Measure the distance it moves before it breaks.

Repeat the experiment with other types of materials. Each material should be of the same length, width, and thickness.

Going Further

Will painting wood or coating it with glue affect elasticity? What additives, if any, will?

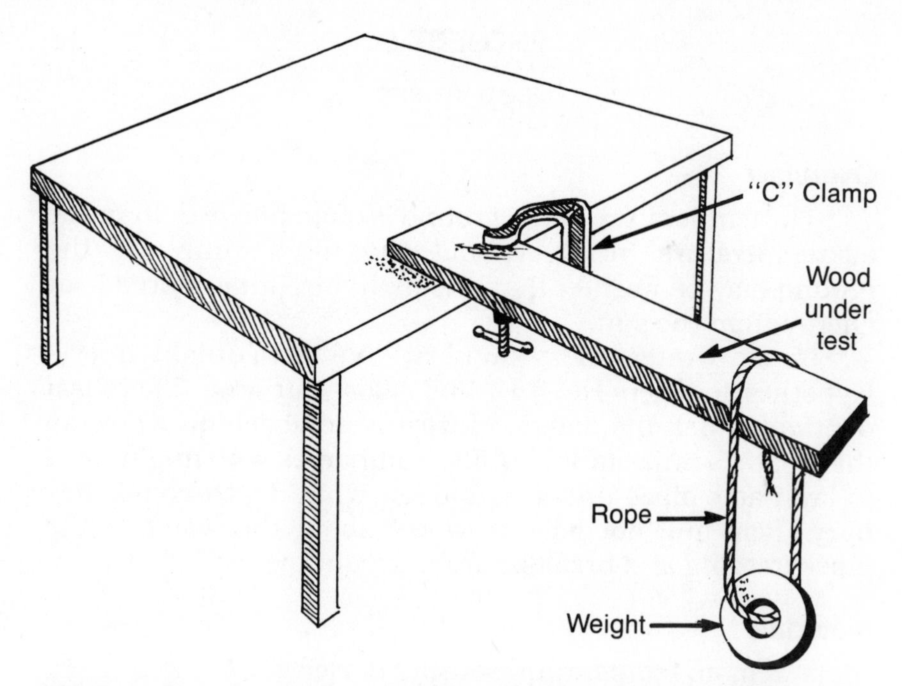

Fig. 2-4. *Testing tensile strength and elasticity in building materials.*

PROJECT 5
Deep Freeze

Overview

The temperature of the ground is often different than the air temperature. In the evening following a sunny day, the ground may be warmer than the night air. It has stored heat energy from the sun.

In the winter, the ground in your area might freeze. Hypothesize where the frost line is for your area. There is a depth at which the danger of frost is nonexistent. Knowing this depth is important to utility companies who might need to lay water pipes underground. At what depth could they bury them and not have to worry about the water in the pipes freezing and breaking from expanding ice?

Materials

- remote temperature-sensing device
- outdoor thermometer
- shovel
- area of ground where you can dig a small hole
- yardstick

Procedure

Obtain a remote sensing thermometer, which has the temperature sensor separate from the indicating device. An excellent example is one of the digital readout home weather instruments, such as the one made by Heathkit Company, Benton Harbor, Michigan 49022.

Dig a one-foot-deep hole in the ground, and place the remote temperature sensor in the hole. Replace the soil back into the hole, covering the sensor. Mount an outdoor thermometer in the air above the buried sensor to read the air temperature.

Every day for a week, record air and ground temperatures, perhaps several times a day as shown in the chart in Fig. 2-5.

At the end of the week, redig the hole to two feet deep and plant the sensor at the two foot mark. Again, keep a temperature log for a week. Repeat the experiment again with the sensor planted three feet deep.

Frost Line Table

TEMP	TIME	MON	TUE	WED	THU	FRI	SAT	SUN
AIR TEMP GND TEMP	7 AM 7 AM							
AIR TEMP GND TEMP	NOON NOON							
AIR TEMP GND TEMP	6 PM 6 PM							
AIR TEMP GND TEMP	10 PM 10 PM							

Fig. 2-5. *Use this chart for recording data to determine where the frost line is. You will need three duplicate charts like this one to collect data for the 1 foot, 2 feet, and 3 feet readings.*

Examine the data you have collected. You might want to draw graphs or charts showing the two temperatures. How do the two temperatures compare? Is one always warmer in the day than in the night? If you carry out your experiment during a cold month, would you be able to conclude the depth at which the frost line is located? If necessary, perform differing depth experiments.

Going Further

Locate different soils within your community. Is there a frost line difference? Test three or more sites.

PROJECT 6

Just Passing Through

Overview

The speed of a seismic wave (an earth vibration) depends on the material through which it must travel. The differences between the materials of the inner core, outer core, mantel, and crust of the earth vary in density and elasticity (its ability to return to the original condition). As a wave passes from one material to another, the energy is refracted (bent). In this project, the transmission of energy through a material will be demonstrated using the height of a wave generated at the other side of the material (see Fig. 2- 6). The initial energy will be supplied by a constant mass from a swing beginning at a constant height. Which material will produce the greatest transmission of energy? Form a hypothesis.

Materials

- cake pan, about 10″ by 14″
- several tongue depressors
- masonry brick (the smallest one you can find)
- one piece of wood cut to the identical size and shape as the brick
- one or two packages of modeling clay (enough to build a clay brick the same size and shape as the masonry brick)
- weight (five to ten ounces)
- string
- adhesive tape
- water
- miscellaneous scraps of wood to construct a swing stand such as the one shown in Fig. 2-7
- centimeter ruler
- tabletop

Procedure

Build the swing stand structure shown in Fig. 2-7. Set up a cake pan with a brick in it. Place it at one end. Tape a tongue depressor to the far end of the pan. Be sure it rests on the bottom. Fill the pan half full with water. Measure the

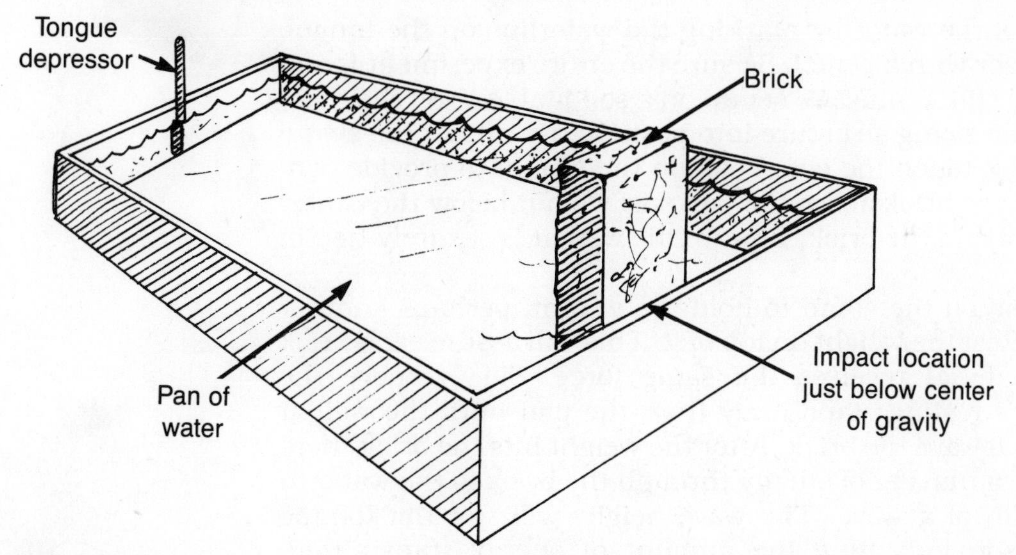

Fig. 2-6. *Construct a tank to test the ability of different materials to transmit seismic waves.*

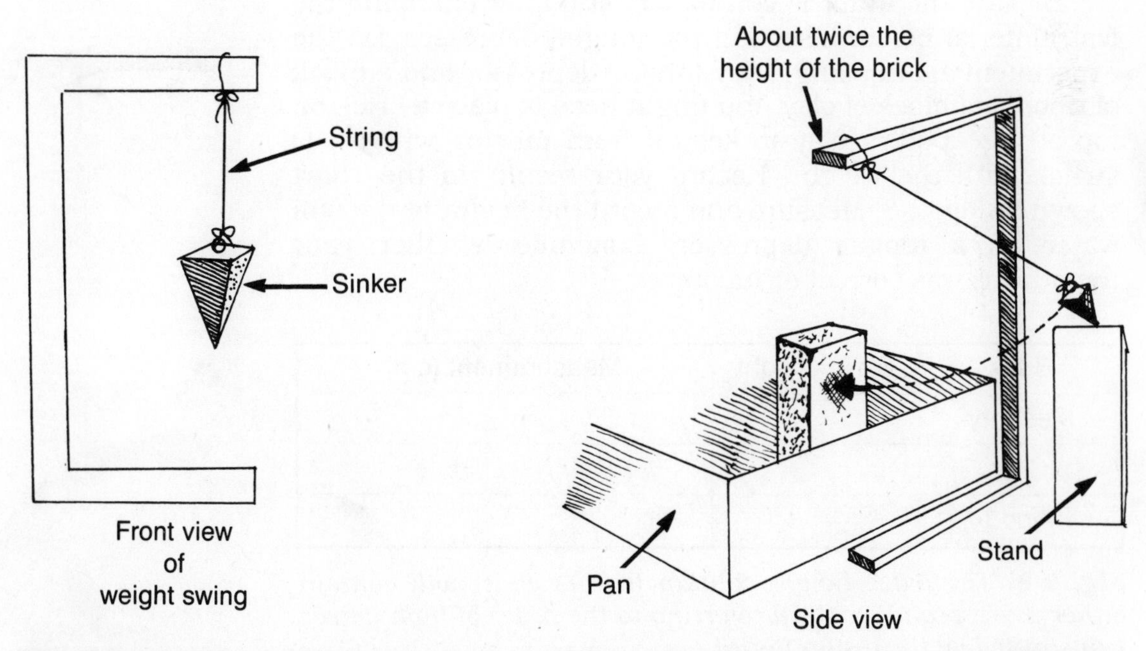

Fig. 2-7. *Construct this structure to swing a weight with equal force each time the experiment is run.*

depth of the water by marking the waterline on the tongue depressor with a pencil. Be sure the entire experiment is stable and still. It must be set up on a solid table or on the floor. Move the swing structure into position. Do not let the swing structure touch the pan. Position the weight to provide contact on the brick above the pan surface but below the center of gravity on the brick. Be sure the weight is securely tied in place.

Position the stand to hold the weight, perhaps another brick. Rest the weight on top of it. This will assure each brick test material receives the same force. Slowly move the weight's resting stand away from the pan until the weight swings toward the brick. After the weight hits the brick there may be a transfer of energy through the brick to the water in the form of a wave. The wave height will wet the tongue depressor, indicating the amount of energy transmitted. Mark and measure the height of the water on the tongue depressor.

Be sure the water is completely still after changing the test material before inserting the tongue depressor. Do the experiment again using a dry tongue depressor and a block of wood and block of clay. You might need to place a brick on top of the wood or clay to keep it from moving when it is struck with the weight. Record your results in the chart shown in Fig. 2-8. Measure and record the height of the still water on a tongue depressor. Conclude whether your hypothesis was correct or incorrect.

Brick	Wave Height	Measurement (cm)
Masonry		
Clay		
Wood		

Fig. 2-8. *The wave height column in this chart will contain either first, second, or third referring to the order of high waves transmitted by the test material.*

Going Further

1. Use different materials (different types of bricks, clays, and woods).
2. Use a glass dish and an overhead projector to view waves on a screen. Determine their frequency.
3. Will dropping straight down (instead of using the swing stand structure) show any difference?
4. Use an outdoor setup around a lake. Try a sledge hammer swing on piling or on a massive boulder. Put a stick in the bottom of the lake for measurement.

PROJECT 7

Seismograph Experiment (Lateral Motion)

Overview

Seismic motion due to earthquakes can produce motion along a line starting from the source of the earthquake (the epicenter). Can a measuring device predict the direction of the source of impact? A device will be constructed and positioned to see whether or not it is possible. The device will be positioned on a table and the table will be hit to cause an impact wave. Then the device will be turned 45 degrees and the impact performed again. Finally, it will be turned another 45 degrees. The shock wave represented by an impact will occur with the same intensity and position in each trial.

Materials

- 1 piece 8″ × 18″ plywood (1/2″ or 3/4″)
- 3/4″ × 4″ × 8″ wood
- 4″ square (to be cut diagonally to make two triangles—these will be used to make corner supports and to support paper roll)
- coffee can (one-pound)
- string (for hanging frame wire)
- five-ounce weight
- dozen marbles
- 18″ molding (1/4″ corner molding)
- two dozen brads
- wood screws
- marking pen
- dowel for paper roll support
- sand or gravel (for weight)
- eye bolt (for adjustment)
- hammer

Procedure

Construct the seismic recording device shown in Fig. 2-9. Place it on a table. There will be three positions, one for each test. Figure 2-10 shows these positions. Impact the table with your hands at point X shown in the diagram. Remove the paper and reattach the string and weight.

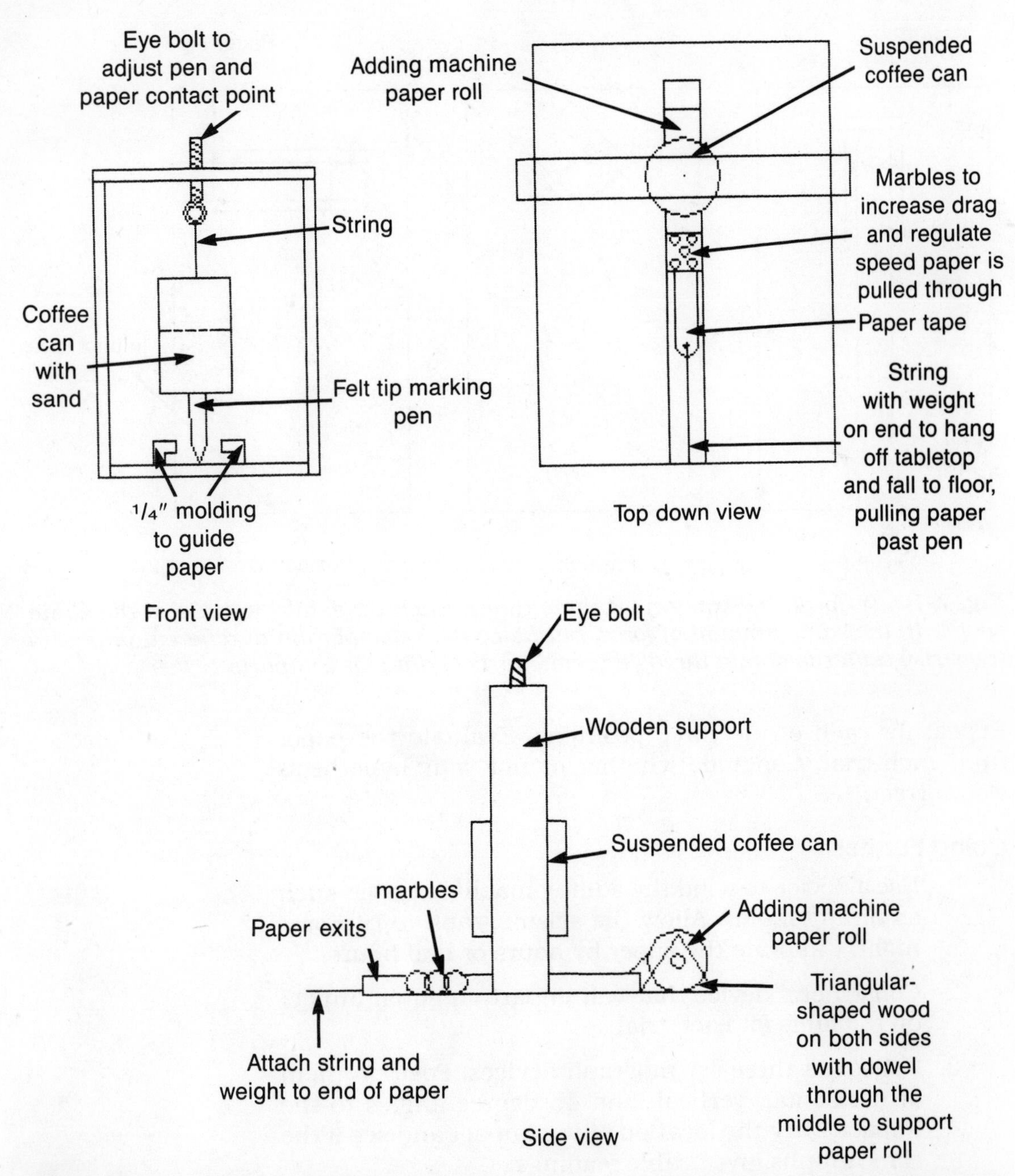

Fig. 2-9. *Construct this seismograph. Paper from an adding machine roll is pulled past a weighted marking pen as the table it rests on is bumped. This records the vibrations. Attach a string and a weight to one end of the paper and let it fall from the table to the ground. This will pull the paper past the recording pen for a second or two, which is all that is needed to obtain data.*

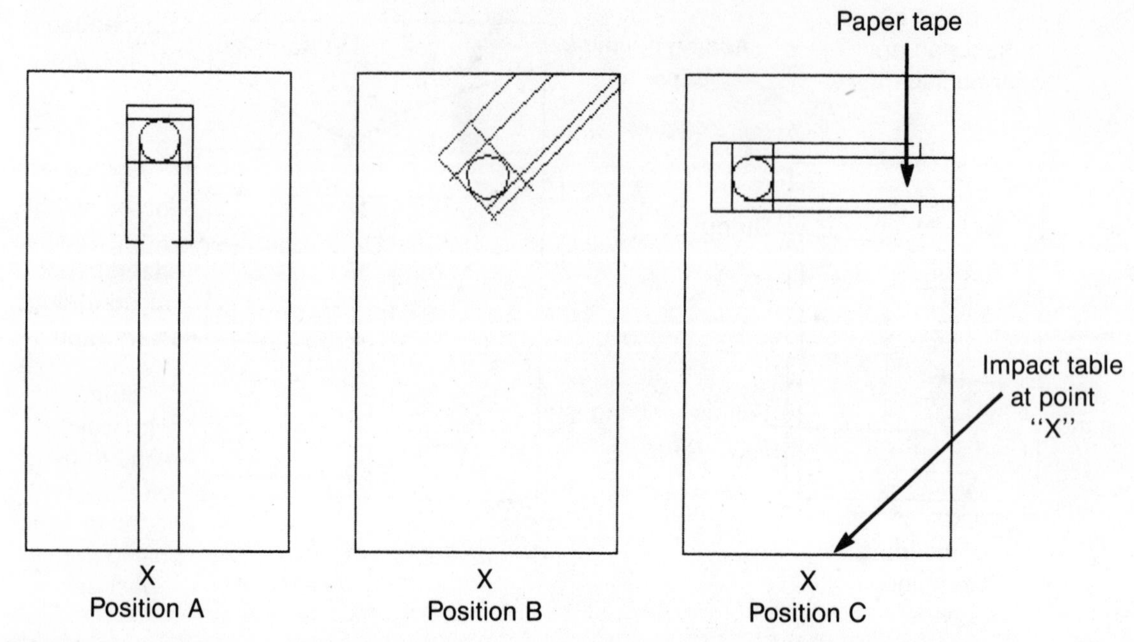

Fig. 2-10. *Perform the experiment three times. Each time, hit the table at the same spot with the same amount of force, but place the seismograph at different angles to the strike center to obtain three different charts of data for comparison.*

Repeat for each of the three positions. Evaluate the paper from each trial. Conclude whether or not your hypothesis was correct.

Going Further

1. Use a device to wind the adding machine paper, such as timing motor. Allow the seismograph to run over night. Calibrate the paper by hours or half hours.
2. Construct a device that will impart an equal impact on the table for each trial.
3. Construct three seismograph devices. Position them at horizontal, vertical, and 45 degree angles to the impact. Vary the location of the impact and see if the three graphs give usable readings.

PROJECT 8

Seismograph Experiment (Vertical Motion)

Overview

The movement of the earth's crust in an up and down motion causes much property damage during an earthquake. Some homes that are built near airports, highways, or factories, also experience motion. Other causes of vibrations are sonic booms, jack hammers, wrecking equipment, pile drivers, and severe storms. Can these stresses be modeled and detected?

Materials

- one piece 8″ × 18″ plywood (1/2″ or 3/4″)
- 3/4″ × 4″ × 8″ wood
- 3″ × 5″ index cards
- coffee can (one pound)
- spring
- five ounce weight
- several inches of molding (1/4″ corner molding to hold 3×5 card in place)
- two dozen brads
- wood screws
- marking pen
- sand or gravel (for weight)
- hammer
- two long dowels (8 inches)
- screw hook

Procedure

Construct the seismograph device shown in Fig. 2-11. Place the device on a table. Run trials to ensure proper functioning. Test by dropping a book on the table to cause vibration. Record your data.

Going Further

Test several different moving vehicles to determine the smoothness of the ride using the device. Form a hypothesis first.

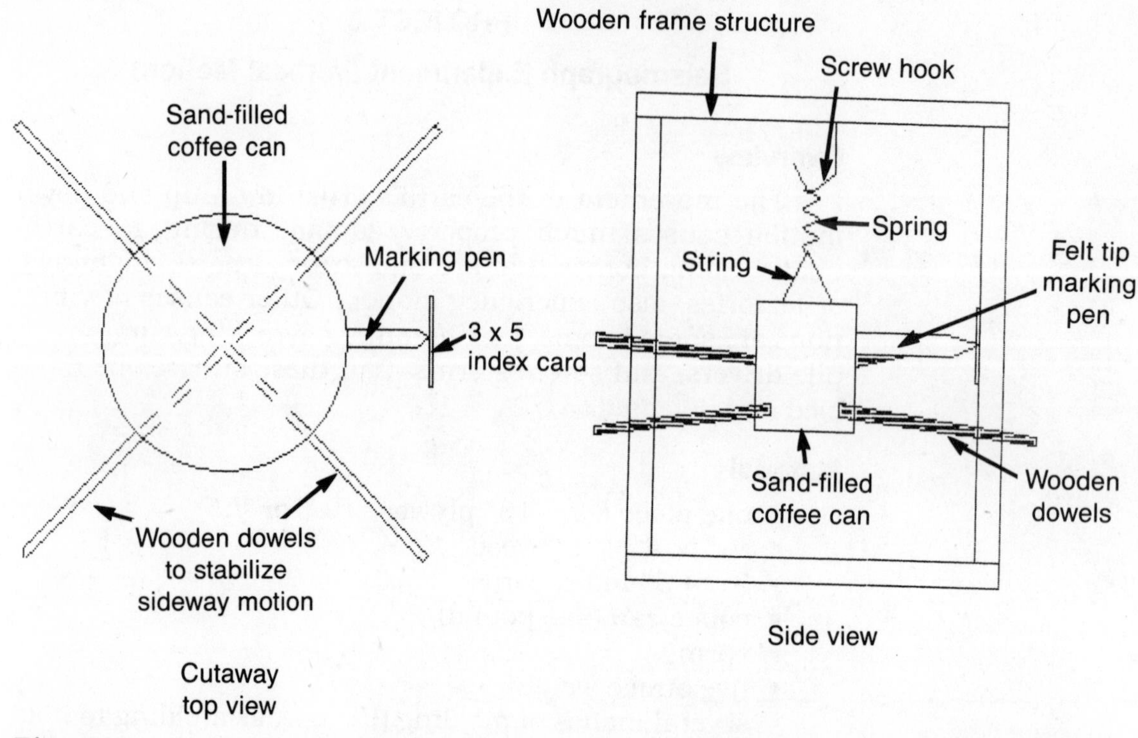

Fig. 2-11. *Construct this seismograph for detecting vertical motion. Secure a 3 x 5 index card to the structure and position it so the marking pen rests at the center of the card. When the table is struck, the card will record the magnitude of the shock wave.*

PROJECT 9

The Magic Lodestone

Overview

Lodestones are naturally occurring magnets. They are iron-bearing materials that have been magnetized due to their position in the earth's crust and its magnetic field. Ancient seafarers used them as compasses. Can we magnetize an iron-bearing rock to make a lodestone? Will one type of material be better suited than another? How can we measure magnetic strength? Gather several different types of iron-bearing rocks and hypothesize which will be the best magnet.

Materials

- several iron-bearing rocks (such as hematite, limonite, magnetite, siderite, taconite)
- iron filings (fragments)
- strong magnet
- balance beam scale or equivalent
- sheet of paper
- small cardboard box (shoe box size)

Procedure

Label the rocks A, B, C, etc. Magnetize rock "A" by using one pole of the strong magnet and stroking the stone gently. After 100 strokes, place "lodestone A" under a piece of clean paper. Shake the iron filings onto the paper (see Fig. 2-12). Next, while holding the paper against the magnetic stone, turn the paper and the rock upside down, allowing some of the iron filings to fall into a catching box such as the one shown in Fig. 2-12. The filings that remain because of the magnetic force can then be dumped onto a scale and measured. Use the same procedure for each rock. Use fresh iron filings for each test to avoid possible magnetized filings. Which one held the most iron filings, thus indicating the strongest magnet? Was your hypothesis correct?

Going Further

1. Use a DC coil (electromagnet) to magnetize the stones. Commercial magnetizers are available.

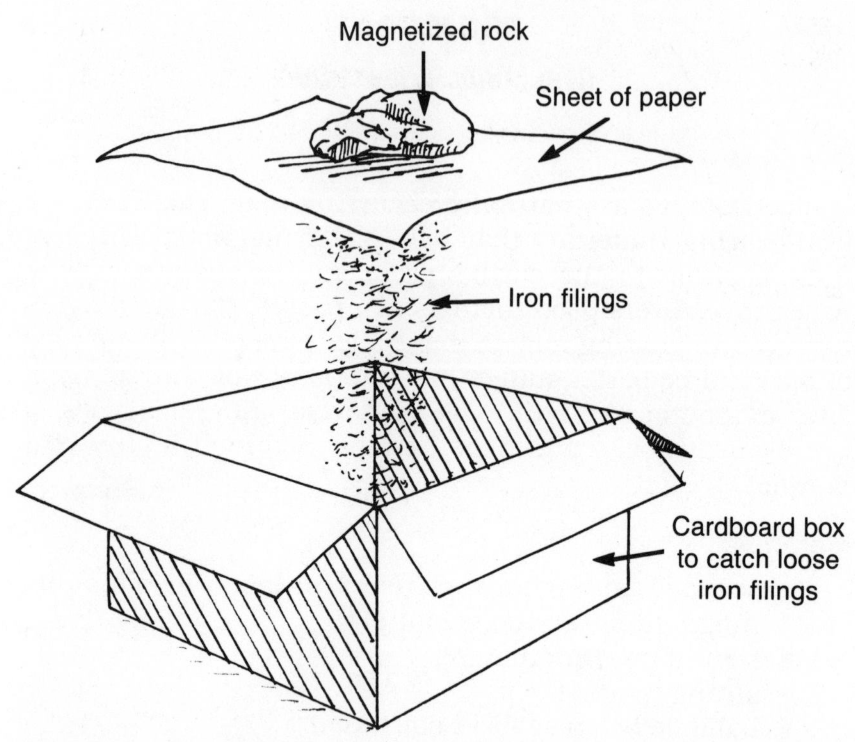

Fig. 2-12. *Sprinkle iron filings onto the sheet of paper resting on top of your homemade magnetic stone. Turn the paper upside down while still holding the stone against the paper. Weigh and compare the iron filings that remain with the results of the other magnetic rocks to see which has the strongest attraction.*

2. Slow moving streams accumulate small quantities of iron. Build your own stone from an iron-bearing sand. Mix with quick drying glue.

PROJECT 10
The Proof is in the Pudding
Adult Supervision Required

Overview

Plate tectonics is the study of the movement of several large segments of the earth's crust which float on top of the astheno (the upper mantel molten material). These plates are slow moving. At some locations, the plates are moving away from each other. At other points they are sliding laterally (sideways) past each other. At the boundaries where plates collide because they are moving toward each other, the earth must release great amounts of built-up pressure. The boundaries of the two colliding plates will respond in one of these ways:

1. The left plate will ride overtop of the right plate, pushing the right plate under it.
2. The right plate will ride overtop of the left plate, pushing the left plate under it.
3. The two plates will push upward, forming mountainlike shapes.
4. The two plates might buckle at many points like an accordion, causing a scallop-shaped ripple in the surrounding ground.

These plate movements are shown in Fig. 2-13.

In this experiment, we will use the crust (skin) on top of a bowl of pudding to simulate the collision of plates. Hypothesize which of the scenarios suggested above will take place when the pudding simulator is used.

Materials

- cake pan (about 8″ × 14″)
- ingredients to make pudding
- a knife (adult supervision)
- two equal-sized wide spatulas

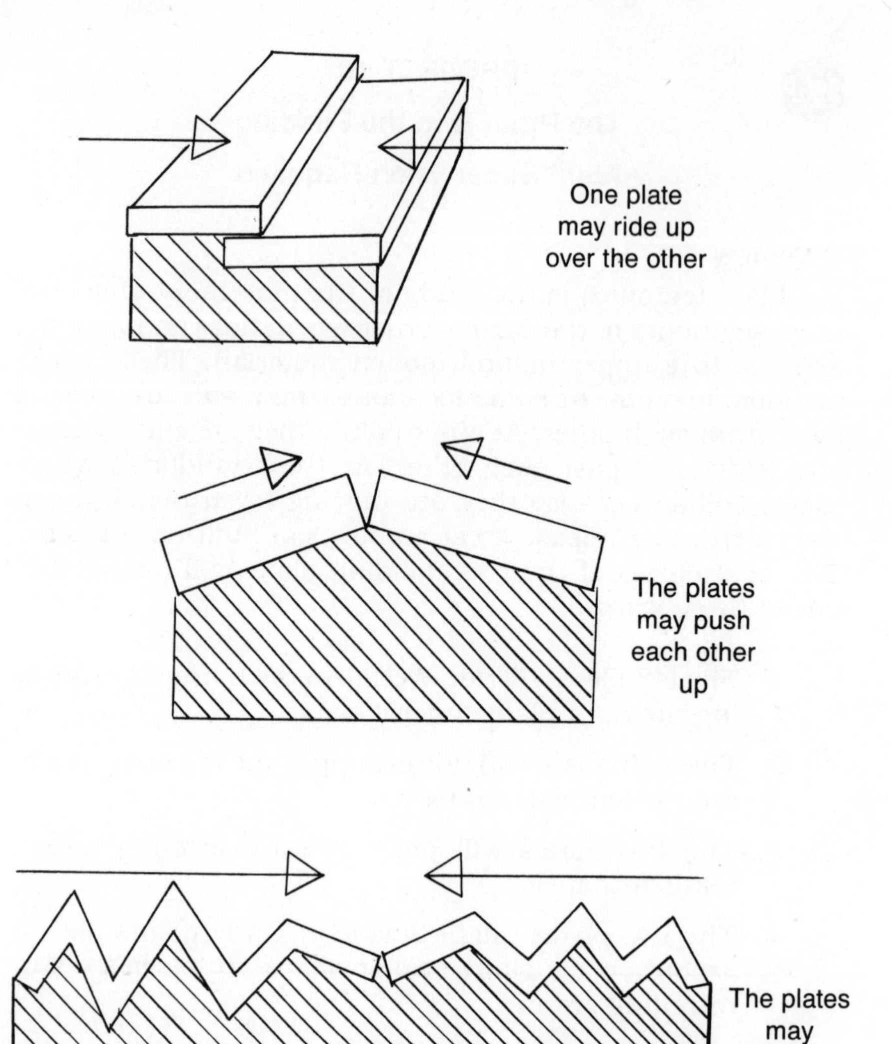

Fig. 2-13. *The possible ways the earth might respond when two plates move toward each other and collide.*

Procedure

Have an adult (caution: hot stove) help make the pudding and fill an 8″ × 14″ cake pan with the pudding. As the pudding cools, a crust will form on the top. Let it set until a crust appears. If the crust forms while the pudding on the bottom is still warm, then this simulates the earth's crust even more accurately because the earth's crustal plates are floating on hot molten material too.

Take a knife and cut all around the outside edges of the cake pan to separate the pudding from the pan. Also make a cut down the middle of the pan, creating two "plates." Put a wide-blade spatula on each of the outside edges of opposite ends and push with slow and equal force toward the center where the middle cut was made. As the two "plates" collide, notice how they move at the colliding boundary and conclude whether or not your hypothesis was correct.

Going Further

Let the pudding set until the skin on top is thicker and repeat the experiment. Are the results the same? Was the internal pudding much cooler?

3

Minerals

The most common solid materials on the earth are minerals. Minerals are inorganic, that is they are not alive nor do they come from living things. They are naturally occurring. Most minerals are made up of one of these elements: oxygen, silicon, aluminum, iron, calcium, sodium, potassium, and magnesium. There are 92 different elements that, combined, make up about 2,500 types of materials. The most abundant mineral on earth is quartz, or beach sand. Combinations of minerals make up rocks.

Each mineral has its own unique set of physical properties. Some of the identifying properties of minerals include color, shape, streak, luster, cleavage, fracture, density, magnetism, and photoluminescence.

Minerals and combinations of minerals have thousands of uses: silicon in electronics, diamond in jewelry, talc in black board chalk, ceramics in tile flooring and insulation, graphite in lubricants.

PROJECT 1
Crystal Clear
Adult Supervision Required

Overview

Crystals are minerals whose atoms are arranged in a pattern that repeats over and over again until the object is large enough to be visible (geometric shapes). Hypothesize that very large crystal structures can be "grown."

Materials

- water
- Pyrex beaker marked in milliliters
- copper sulfate in crystalline form
- spoon
- string or thread
- pencil or popsicle stick

Procedure

Have an adult boil some water. Pour 50 milliliters of hot water into a pyrex beaker. Slowly add the crystal mineral, copper sulfate ($CuSO_4$). Continue adding copper sulfate to the boiling water, and stir until crystal particles begin settling on the bottom. At this point, the solution has reached a supersaturated condition where no more mineral can be dissolved in the water.

Let it cool and stand at room temperature overnight. Do not disturb it. Put it somewhere, such as on a window sill, where it will not get bumped. Within the next two or three days, remove the biggest single crystal you can find on the bottom of the beaker. This will be used as a "seed" crystal upon which we will attempt to collect more crystals and build a bigger one. Set the seed crystal aside.

Boil the solution in the beaker and again add more mineral until the solution becomes supersaturated. We do not want the seed crystal to be in the beaker at this time, because the hot water might dissolve it.

When the solution reaches a supersaturated condition, remove it from the stove. Let it stand until it cools to room temperature. Tie a piece of string around the large seed crystal and tie the other end to a pencil or popsicle stick. Put the

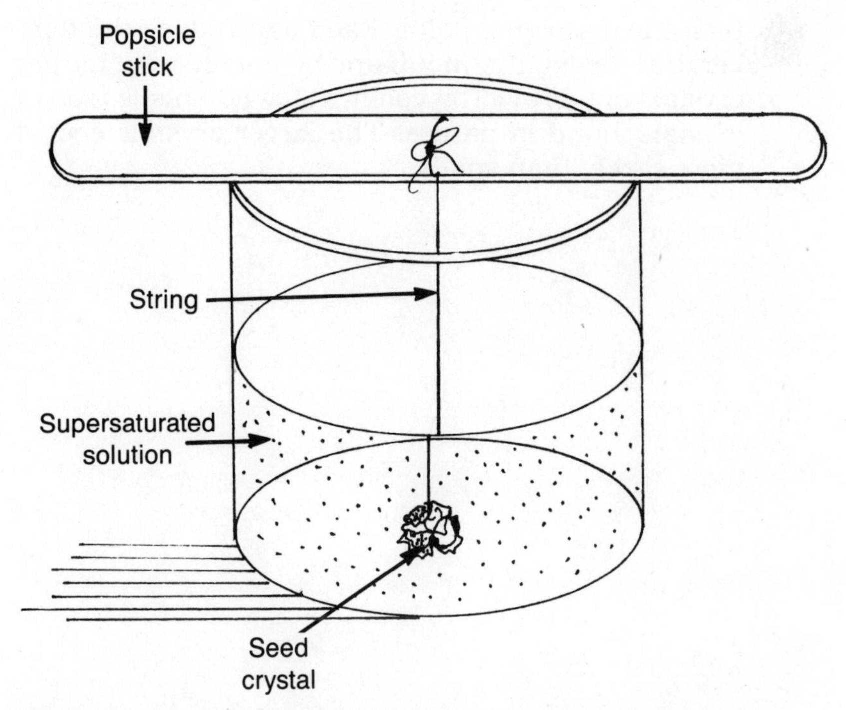

Fig. 3-1. *Crystal structures can be "grown" to large sizes.*

pencil across the top of the beaker and let the seed crystal hang down into the solution (see Fig. 3-1). Again, let the solution stand undisturbed overnight. After several days, inspect the seed crystal and conclude as to whether or not your hypothesis was correct.

Going Further

1. Repeat the experiment, but instead of using copper sulfate, use sugar. Sugar is organic and not a mineral, but it can be used to demonstrate the behavior of growing crystals. When the experiment is over, the sugar crystals can be eaten.

2. Repeat the experiment using other crystal-type minerals, such as salt, and hypothesize which ones can be grown to form the largest structures.

3. Find the temperature at which supersulphate (or other mineral you choose) dissolves. Place the suspended seed crystal in a supersaturated solution in an incubator. Set the temperature one degree lower

than the dissolving point. Each day, reduce the temperature inside the incubator by one degree. Larger crystals are formed by cooling slowly. This is true of crystals found in nature. The larger crystals cooled more slowly than smaller ones.

PROJECT 2
Salt in the Wound

Overview

Salt (sodium chloride) causes water to freeze at a lower temperature. It is used in ice cream makers to obtain lower temperatures. Also, it is used to melt the ice from road surfaces and sidewalks. How cold can water be? Tap water freezes at 0 degrees Celsius (32 degrees Fahrenheit). What temperature can be obtained in seawater before it freezes? Can a solution containing table salt be made to obtain a lowest possible temperature? How much salt must be added to a specific quantity of water to decrease the freezing point to the lowest temperature, beyond which additional salt will not cause a lower temperature? Form a hypothesis.

Materials

- several, plastic 8-ounce cups
- measuring cup
- box of table salt
- freezer
- two thermometers
- teaspoon

Procedure

Pour six ounces of tap water into two, eight-ounce cups. Add a teaspoon of salt to one cup and two teaspoons of salt to the other. Stir to completely dissolve the salt in the water. Insert a small thermometer in each cup. Place the cups in a freezer. Periodically check them to see if any ice has started to form. When ice first appears as a rime (thin layer) on the surface, record the temperature.

Repeat the experiment, each time dissolving more salt in the water, until you eventually hit the point of saturation. Saturation occurs when no more salt can be dissolved in the water. Adding more salt merely settles it in the bottom of the cup as crystals. Use a chart similar to the one shown in Fig. 3-2 to record your data. Was your hypothesis correct?

Salt in the Wound Table

Cup Number	Teaspoons of Salt Added	Freezing Temperature

Fig. 3-2. *Use this chart for recording data.*

Going Further

1. Use a different kind of salt such as potassium chloride. Research the various types of salt.
2. Determine by experiment the freezing point of seawater or water from a lake (which has other substances in it besides H_2O). How does this temperature compare to the ones obtained in your experiment above?

PROJECT 3
In Hot Water
Adult Supervision Required

Overview

Water boils at 212 degrees Fahrenheit (100 degrees Celsius) and will get no hotter. At boiling, the liquid becomes a gas, or steam. A pressure cooker allows water to become hotter than 212 degrees F before boiling. This device causes an increase in pressure. Can we find another method for increasing the boiling point? Will the addition of salt change the boiling point? Will adding more salt matter? How about a supersaturated saltwater solution? Form a hypothesis.

Materials

- table salt
- water
- stove burner
- thermometer (must read up to 150 degrees Celsius or 300 degrees Fahrenheit)
- a two-quart cooking pot with a lip around the top
- two clothes pins (spring clip type)
- measuring cup and spoon

Procedure

Pour two cups of water into a pot. Measure and dissolve as much salt as possible in the water (at room temperature). Write down how much salt you added (teaspoons, tablespoons, grams). Next, add two more cups of water and stir it (this makes a 50 percent diluted solution). As shown in Fig. 3-3, use clothes pins to suspend a thermometer in the pot of water (the thermometer should not touch the bottom of the pot). Heat the solution and measure the increasing temperature. Adult supervision is required when working around a hot stove. Record the highest temperature reached.

Next, add an amount of salt equal to the initial quantity put in. This makes a saturated solution at room temperature. Then heat the solution to its hottest temperature. Add as much more salt as will dissolve. The solution is now supersaturated. Measure and record your results, and conclude whether your hypothesis was correct.

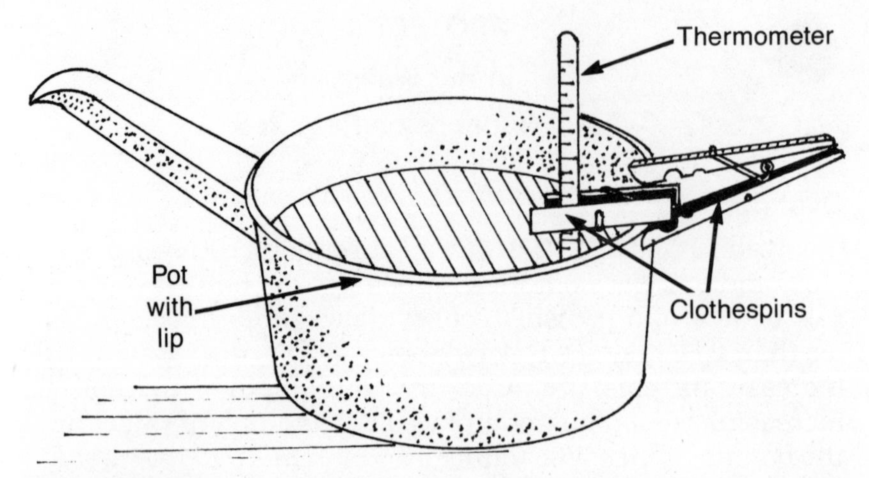

Fig. 3-3. *Use two clothespins to suspend a thermometer in the boiling water solution. Don't let the thermometer rest on the bottom of the pot.*

Going Further

1. Use seawater instead of saltwater in the above experiment. Seawater contains many different salts.
2. Use a layer of cooking oil instead of saltwater in the above experiment. Will the oil float on top and cause the water to become hotter than 212 degrees Fahrenheit? Use a 500 ml beaker with 200 ml of water and 200 ml of cooking oil as a top layer.
3. Use a pressure cooker with the salt or cooking oil experiments above.
4. Would sugar work as well as salt? Would it work better?

PROJECT 4
Rock Garden

Overview

Rocks have many identifying characteristics. Collect and identify rocks found in your neighborhood. How many different characteristics can you use in identification? Experiment to discover information. How do local rocks compare to other rocks in your state? How did the rocks get in your neighborhood? Some people import rocks from far away places to be used in driveways and landscaping.

Materials

- several egg cartons
- research books on rocks
- paper and pencils
- collection of local rocks

Procedure

Collect as many different rock specimens as you can from around your neighborhood. Use empty egg cartons to hold the specimens. Examine the rocks closely. Assign each rock a specimen number and label it. Design a chart such as the one shown in Fig. 3-4, listing various properties. Consider the color (coal is black, sulphur is yellow). Give rocks the "heft" test. Place a rock in your palm and move it up and down to get a sense of weight. Which ones of similar size and shape are heavier? It is best to compare two rocks by holding one at a time in your right hand or left hand, but not one in each hand, because one arm might be stronger than the other and make it seem like one rock is lighter. How could you measure this more accurately? Investigate specific gravity, the density of an object compared to the density of another object.

Compare the rocks by feel: smooth, rough, oily. Test them for specific gravity, crystalline structure, luster, streak color, hardness, magnetic properties, and photoluminescence, the luminescence caused by the absorption of infrared radiation, visible light, or ultraviolet light. Using a rock identification book, attempt to name all of the specimens based on the data you placed on your chart.

Rock Garden Table

Rock	Specimen #1	Specimen #2	Specimen #3
Feel			
Specific Gravity			
Crystaline Structure			
Luster			
Streak			
Color			
Hardness			
Magnetic Property			
Photolum-inesence			

Fig. 3-4. *Use this chart for recording data.*

Going Further

Can you research information on the rocks? Were they produced by industry or used in landscaping? Are there any distinguishing features such as fossil materials?

PROJECT 5
Building Up or Down

Overview

There are natural structures that form by accretion (the slow, steady buildup of material). Stalactites and stalagmites form in caves where water containing minerals seeps through and drips. The minerals adhere (stick) to other molecules of the same substance and increase the size of the hanging stalactite. Most often the mineral is calcite in limestone caves. The cave floor becomes spattered with the dripping solution and the "growth" of a stalagmite occurs. Over a long period of time the two pieces lengthen and meet forming a column. Can an artificial situation simulate these formations? Which materials would be best? Would a plaster of paris solution form a stalagmite better than fine sand carried by a solution of water-soluble (capable of dissolving in water) glue? Form a hypothesis.

Materials

- two aluminum cake pans
- a box of plaster of paris (be sure to read warnings on the label)
- fine sand (equal amount to the plaster of paris)
- two, empty one-gallon plastic milk jugs
- 2″ × 2″ × 8″ wooden blocks
- eight ounce bottle of water-soluble glue (Elmers)
- two ring stands or homemade stands
- one washcloth, cut into four strips
- water
- string

Procedure

Set up the two cake pans with one end on a block. This will permit the extra material to run down to the low end and be reused.

Mix the solutions in separate containers (the milk jugs). In one jug, make a solution of plaster of paris. In the other, mix the water, glue, and fine sand solution. Keep these stock solutions wet to keep them in a liquid state. Water can be added. Cover when not in use.

Position the ring stands over each pan. Saturate one of the washcloth strips in the plaster solution and the other in the glue solution. Next, tie them to the rings on the ring stands to allow them to drip down onto the pan (see Fig. 3-5). Do not make the height of the strips more than two or three

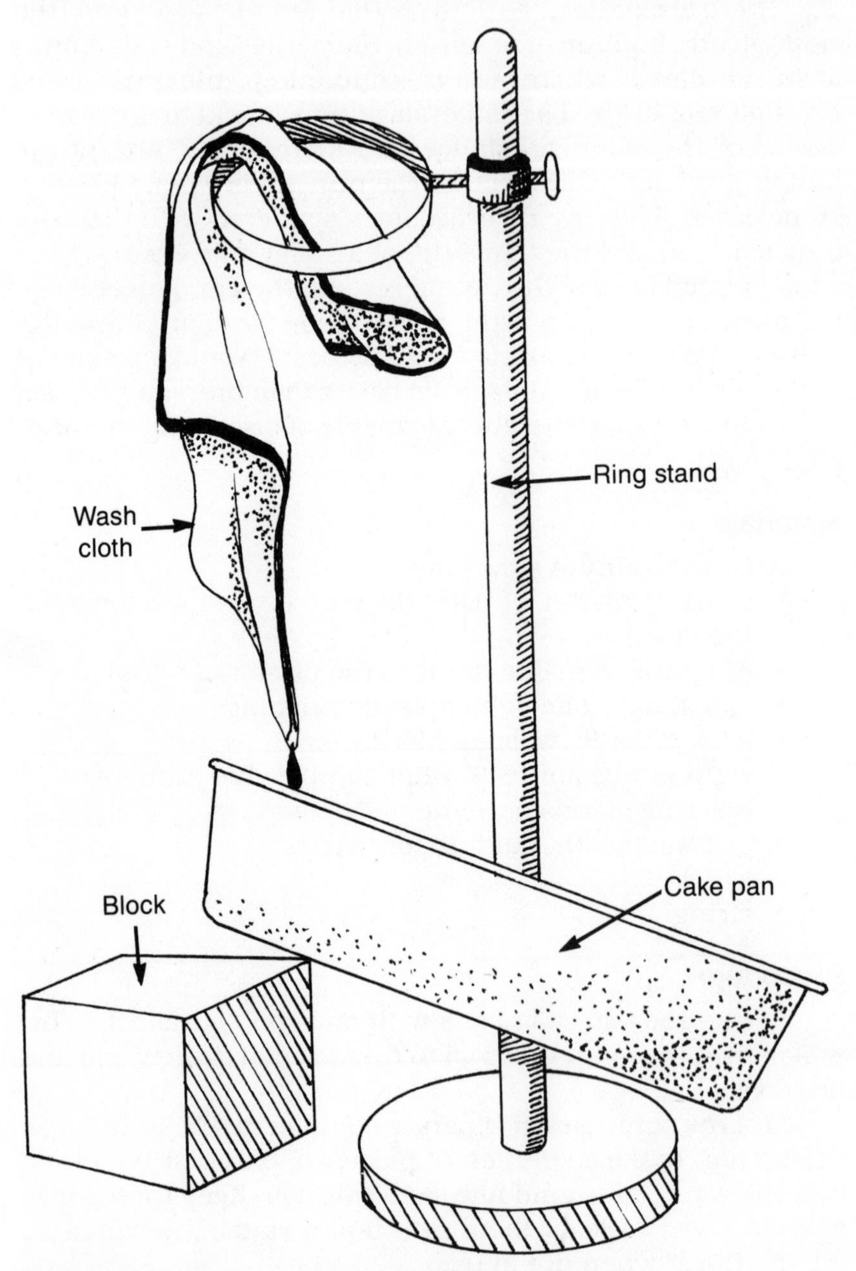

Fig. 3-5. *Build a drip system to form stalagmites.*

inches above the pan. From this height they will not splash much. When they stop dripping, resaturate the cloths and repeat over and over. Do this for several days, morning and night. Measure the results. Record your data and conclude whether your hypothesis was correct.

Going Further

1. Establish a ceiling for the simulation of stalactite formation. Attempt to form a column.
2. Try other materials for column buildup: wall paper paste, sugar water, salt water, batter (pancake).

PROJECT 6

What You See Is *Not* What You Get

Overview

When examining a mineral, perhaps the most obvious characteristic that stands out is color. The outside color, however, might not be the same color under the surface. It is well known that iron produces rust. Rust is a combination of the iron material with oxygen. Copper oxidizes to form a layer of green material. Aluminum oxidizes and forms a white powder. Hypothesize what colors will streak when six sample rock specimens are tested.

Materials

- Mohs scale of hardness test kit (See the Resource List for suggested suppliers of science test kits.)
- six, unknown mineral specimens, with a hardness of less than five on the Mohs scale
- one, unglazed ceramic tile (obtainable at a pottery or hobby shop, or sometimes supplied with the purchase of a Mohs test kit)
- tape or glue
- pen

Procedure

Collect six rock specimens. The streak test is done by rubbing a rock against an unglazed ceramic tile, which has an index of approximately 5 on the Mohs scale of hardness. Therefore the rocks must be softer than the ceramic tile. Record and identify each specimen by giving them a number. To do this, tape or paste a small dot with a number on it to the specimen. Record the observed color (outside appearance) for each and record in a chart such as the one shown in Fig. 3-6. Perform a Mohs scale of hardness test on each rock. To streak test, mark the tile with the specimen by rubbing it hard against the tile. Record the color results. Clean the tile after each use by wiping it.

When all of the tests are completed, use an identification book and list the four most likely possibilities for each speci-

Streak Test Table

Specimen	Color	Streak	Hardness	Possibilities
#1				1. 2. 3. 4.
#2				1. 2. 3. 4.
#3				1. 2. 3. 4.
#4				1. 2. 3. 4.
#5				1. 2. 3. 4.
#6				1. 2. 3. 4.

Fig. 3-6. *Use this chart for recording data.*

men. Are there other tests that will help narrow the choices down? Can you do them? Was your hypothesis correct?

Going Further

Demonstrate the concept of streak testing by using a rusty nail, an oxidized penny, and a tarnished piece of silver. For example, an oxidized penny looks green in color but a streak test using a file will reveal the underlying copper color. Provide other examples of oxidized materials.

4

Rocks

Rocks make up the solid part of the earth's surface. They are combinations of one or more minerals. Rock formations are found under layers of soil and beneath the oceans. When rock is broken down into tiny bits, it can mix with organic material (decomposing animal and plant life) to form soil.

Rocks have many characteristics. They vary in chemical composition, color, hardness, magnetic properties, and shape. The characteristics can be used to help identify them.

There are many valuable rocks that mankind has put to good use. Granite and marble are used to construct buildings. Highway roads can be made firm by laying a foundation of rock. Concrete, made from various crushed rocks, is used in everything from building sidewalks to dams. Radioactive ores, such as uranium, are useful in the field of medicine and electricity generation. Hard rocks, such as diamonds, are valuable cutting tools. You probably listen to your favorite record with the help of a diamond or sapphire record player needle. And of course, gems have been treasured down through the ages for their beauty.

Rocks are classified by the way in which they were formed. Igneous rocks were formed by a cooling process, such as hot lava from a volcano. Sedimentary rocks are made by a layering of materials that settle. Metamorphic rocks were once either igneous or sedimentary rocks that have changed by heat and pressure. Rocks are ever-changing in a cycle.

PROJECT 1
Building Your Own Building

Overview

Concrete is a man-made building material. Ancient buildings were built from large blocks of rock cut from huge natural formations and transported to the site of the building. To shape them, a stone cutter had to chisel each block one at a time. Today, we have concrete, cinder block, steel girders, and brick. In this project, we will make various consistencies of concrete.

Will one of our test mixes be stronger than the others? Will the strongest be the heaviest? The heaviest will be determined by measuring the specific gravity of the sample. Specific gravity is a comparison between an unknown and an equal volume of water by weight (the procedure section that follows will offer further details).

Materials

- concrete mix (Redi-mix)
- sand
- water
- gravel
- stirrer (wooden dowel)
- six-ounce paper cups
- scale (balance beam can be homemade)
- weights (measuring)
- measuring cup with pour spout (4 cup)
- bowl or cup (to catch the overflow of displaced water)
- weights
- tablespoon (measuring tablespoon)

Procedure

Set up six paper cups to receive different mixes. Be sure to identify each cup with a label or marking. Add enough water to thoroughly mix and make a smooth consistency. Try to make the consistency the same for each mixture. Measure the quantity of water used (in tablespoons) for each mixture, and record it on a chart such as the one shown in Fig. 4-1. Be sure to mix the entire quantity of materials completely. There should be no dry material in the bottom or on

Building Your Own Building Table
Table 1 of 2

All measurements are in tablespoons				
Cup #	Sand	Water Used	Redi-mix	Gravel
1	6		6	
2			4	6
3	4		6	4
4	3		6	3
5	9		3	
6			3	9

Fig. 4-1. *Chart one of two for recording data.*

the sides. Mark the number in the surface as it hardens. Allow one week for it to dry. Then remove the paper cups and dry for another week.

Before testing for strength, determine which piece is the most dense or has the greatest specific gravity and what the specific gravity will be.

To measure for specific gravity, weigh stone number one on a scale. Record its weight in the chart shown in Fig. 4-2. Weigh a dry bowl or cup that will be used to catch displaced, overflowing water. Fill a measuring cup to overflowing with water. When the water stops flowing, place the dry catch bowl or cup in position under the measuring cup's spout. Gently place the stone in the water filled cup, causing some water to overflow out the spout and into the catch cup. When it stops flowing, weigh the catch cup with the water. Subtract the weight of the catch cup to find the weight of the water. Record the weight of the water on the chart. Divide the weight of the water into the weight of the stone (make it accurate to two decimal places) to arrive at a figure for specific gravity. Record these numbers on the chart. Follow the same procedure for each stone. Be sure to dry the catch cup before each test.

Experiment to determine which stone can support the most weight. Add weights on top of them until they break. Hypothesize which will be stronger. Will there be any difference? Conclude whether your hypothesis was correct.

Building Your Own Building
Table 2 of 2

Stone Number	Stone Weight	Catch Cup and Displaced Water	Weight of Catch Cup	Weight of Displaced Water	Specific Gravity
1					
2					
3					
4					
5					
6					

Fig. 4-2. *Chart two of two for recording data.*

Going Further

1. Use glue instead of concrete mix. Make blocks using milk carton molds.
2. Perform a cost analysis for each mix. If a job required 7 cubic yards, which would be cheapest?
3. Which sample weathers better?
4. Test for impact (wear safety goggles). Drop a weight from a measured height onto the brick. Increase the height until shattering occurs.

PROJECT 2
Castles Made of Sand

Overview

Silicon is the most abundant mineral in the earth's crust. Oxygen is the most abundant gas. Silicon plus oxygen is silicate. Silicate is sand. It is a key ingredient in the manufacturing of building construction materials. In this project, sand is used with different types of glue to make building materials. The materials will be tested for strength.

Before you begin, select three different types of glue and hypothesize which will make the strongest building materials when mixed with sand.

Materials

- measuring cup
- teaspoon
- sand
- three different water-soluble types of glue (such as model airplane glue, Elmer's glue)
- paper plates
- water
- C-clamp
- vise grip pliers
- weights
- string or small rope

Procedure

You will make three pancakelike "brick" material samples, each less than one-half inch thick. Use paper plates to mix the ingredients and to act as a mold for each sample. All three mixtures must use the same amount of sand, water, and glue so that the only variable in the experiment is the type of glue. Determine that each type of glue is water-soluble.

Make a pancake brick by mixing two ounces of water, one ounce of glue (pick one type), and five ounces of sand (measured in ounces by volume, not by weight). Stir to make an even consistency. The thickness of your brick should not be more than one-half inch thick. Mix two more brick sam-

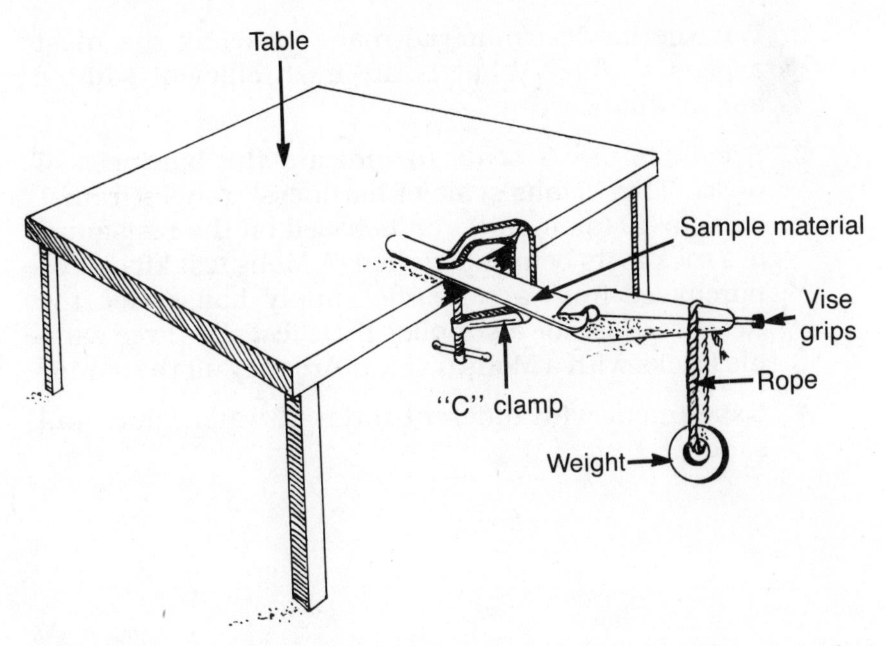

Fig. 4-3. *Setup for testing the relative strength of homemade building materials.*

ples, each using a different type of glue. Allow one week to set, dry, and harden.

Now the three samples can be tested for tensile strength. Using one at a time, secure one end of a brick to a table using a C-clamp. Clamp a pair of vise grip pliers to the other end. Tie a piece of string or rope to the end of the vise grips (see Fig. 4-3). Begin hanging weights on the end of the rope until the sample brick breaks. Record the total weight needed to break a piece off. Repeat the process with the other samples.

Which sample was the strongest? Does it have any value as a building material? Conclude as to whether or not your hypothesis was correct.

Going Further

1. Test samples for impact strength by dropping weights, such as fishing sinkers onto it from equal heights (wear safety goggles). Some materials, such as carbide steel-cutting tools, are very strong in one direction but can shatter easily when struck broadside.

2. Was the hardest material made by using the most expensive glue? What is the most efficient sample and at what cost?
3. Scientists use a scale to indicate the hardness of rocks. This "Mohs scale of hardness" ranges from 1 (talc) to 10 (diamond) and is based on the resistance of a rock to its being scratched. A Mohs test kit can be purchased from a scientific supply house (see the Resource List for a supplier list). Test the three sample bricks with a Mohs test kit. Are they all the same?
4. Experiment with different ratios of water, glue, and sand.

PROJECT 3
The Bigger the Better?

Overview

Sand has been used for ages in the construction of building materials. Does the size of the sand particles used in the mix affect strength? In this project, we will construct three sample "bricks," each using different size sand particles (small, medium, and large). Hypothesize which brick you believe will be the strongest. Will the largest particles increase strength?

Materials

- measuring cup
- teaspoon
- sand
- Elmer's glue
- paper plates
- water
- C-clamp
- vise grip pliers
- weights
- string or small rope
- soil sieves (with three different size screen openings)

Procedure

You will make three pancakelike "brick" material samples, each less than one-half inch thick. Paper plates will be used to mix the ingredients and to act as a mold for each sample. All three mixtures must use the same amount of sand, water, and glue so that the only variable in the experiment is the size of the sand particles (small, medium, and large).

Using soil sieves, sift sand to make three piles of sand particles arranged by size.

Make a pancake brick by mixing two ounces of water, one ounce of glue, and five ounces of sand (measured in ounces by volume, not weight). Stir to make an even consistency. The thickness of your brick should not be more than one-half inch thick.

Mix two more brick samples, each using a different size sand particle. Allow one week to set, dry, and harden.

Now the three samples can be tested for tensile strength. Using one brick at a time, secure one end of the brick to a table with a C-clamp. Clamp a pair of vise grip pliers to the other end. Tie a piece of string or rope to the end of the vise grips (see Fig. 4-3 in the last experiment, *Castles Made of Sand*). Begin hanging weights on the end of the rope until the sample brick breaks. Record the total weight needed to break a piece off. Be sure to position the C-clamp and pliers similarly in each test.

Which sample was the strongest? Does it have any value as a building material? Conclude whether your hypothesis was correct.

Going Further

1. Test samples for impact strength by dropping weights, such as fishing sinkers, from equal heights (wear safety goggles). Some materials, such as carbide steel-cutting tools, are very strong in one direction but can easily shatter when struck broadside.

2. Scientists use a scale to indicate the hardness of rocks. This "Mohs scale of hardness" ranges from 1 (talc) to 10 (diamond) and is based on the resistance of a rock to being scratched. A Mohs test kit can be purchased from a scientific supply house (see the Resource List for a supplier list). Test the three sample bricks with a Mohs test kit.

PROJECT 4
Can You Feel the Difference?

Overview

Sandstone is an accumulated material. It forms by layering and compacting. Minerals carried by water help cement the small particles to form stone. As the sandstone gets pushed lower and lower into the earth's crust, the pressure and heat increase to form shale. Shale then heats and melts as it gets pushed lower into the earth's crust. When it cools it forms granite. Can granite, shale, and sandstone be separated by texture (how it feels)? Hypothesize the results of testing people as they examine the texture of these rocks by touch alone.

Materials

- one piece of sandstone, at least two inches square
- one piece of shale, at least two inches square
- one piece of granite, at least two inches square
- small cardboard box
- dark cloth to cover the box

Procedure

Set up a box with a cover (see Fig. 4-4). Place all three rock specimens in the box. Ask individuals (people) to test the texture by putting their hand into the box (behind the curtain) and arranging the rocks by texture. Tell them to put the smoothest to the left and the most coarse to the right. Check the results of each person's test and log the data. Will each test subject place the rocks correctly? Test at least ten or more people. Conclude whether your hypothesis was correct.

Going Further

What other specimens can be identified by their texture? Talc is oily. Mica is smooth. There are different types of coal.

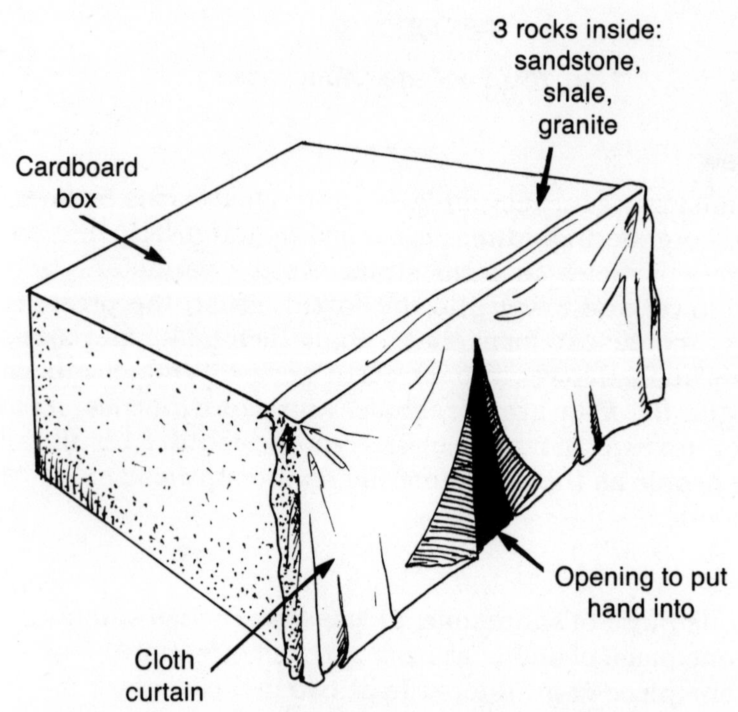

Fig. 4-4. *Use a cloth as a curtain to keep people from seeing the rocks inside. Have people place them in order of texture, smoothest to roughest.*

PROJECT 5

I Tumble for You

Overview

Scientists use a scale to indicate the hardness of rocks. The Mohs scale of hardness ranges from 1 (talc) to 10 (diamond) and is based on the resistance of a rock to its being scratched. Diamond is the hardest natural-forming material.

A rock tumbler uses sand to smooth rocks by abrasion. Hypothesize that adding particles of greater hardness (having a higher Mohs scale number) will shorten the time a rock must remain in a tumbler to be made smooth. This saves time, energy, and frees the tumbler up for other work.

Materials

- rock tumbler
- scrapings from sandpaper (fold it and rub it against itself, collecting particles which fall off)
- scrapings from corundum paper
- ten common local rocks
- Mohs hardness test kit (available at scientific supply firms—see the Resource List)

Procedure

Collect ten common rocks found in your area. They should be about equal in size and of the same hardness. Pair the rocks that are the same type and that have approximately the same number of sharp edges and corners (see Fig. 4-5).

Scrape particles from sandpaper and corundum paper. Corundum is extremely hard.

Place the particles from the sandpaper into a rock tumbler along with a pair of rocks, tumble them for several days. Observe how much abrasion has taken place. The rocks should have the same hardness. Add water to the tumbler also.

Now place a similar pair of rocks in the tumbler for the same amount of time but use corundum particles instead of sand. Do this with all pairs of rocks. Compare the results and conclude whether your hypothesis was correct.

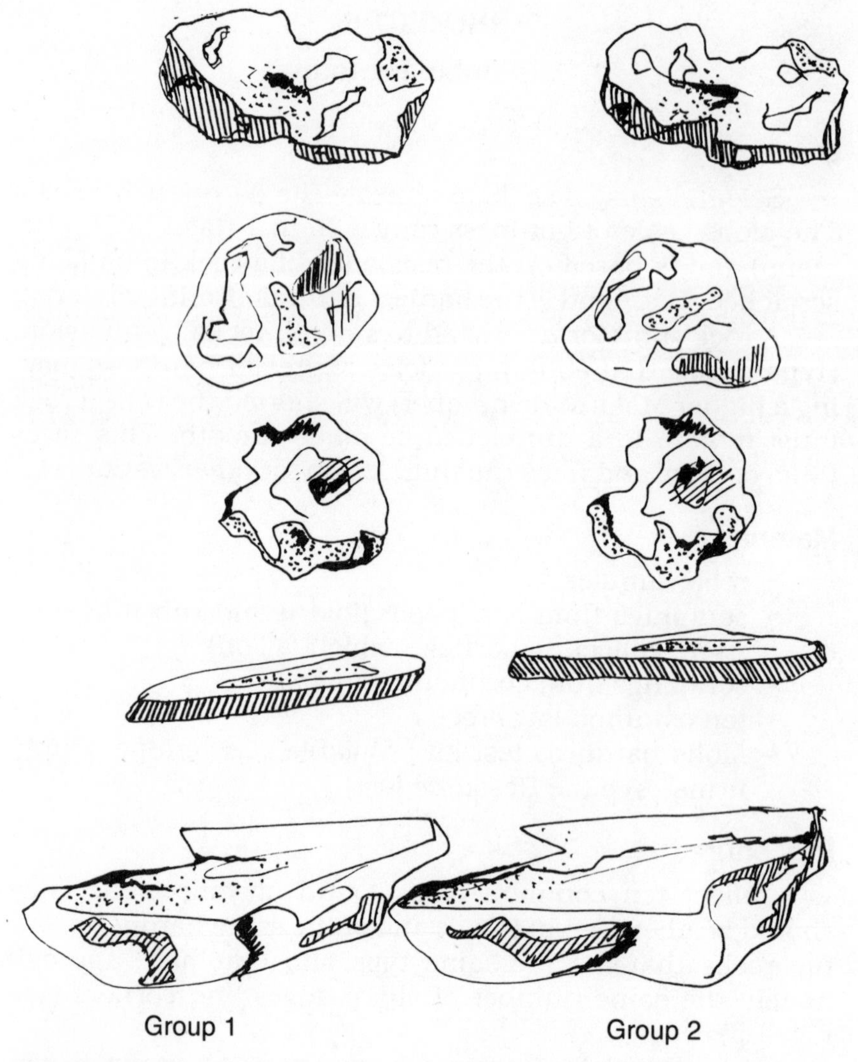

Fig. 4-5. *Pair rocks into two groups based on their type and sharp edge features.*

Going Further

Make your own tumbler material for cleaning metal jewelry.

PROJECT 6
The Bubbling Answer
Adult Supervision Required

Overview

The compound calcium carbonate is present in bones, shells, limestone, and other rocks. It reacts with a diluted sulfuric acid solution of fifteen percent to form bubbles. Limestone can be found in caves and in conglomerates that include shell material. The White Cliffs of Dover in England are limestone. Do locally gathered materials have calcium carbonate present?

Materials

- 15% diluted sulfuric acid solution, appx. 20 milliliters
- bucket
- water
- rubber gloves (safety for handling acid)
- eye dropper
- various local rock specimens
- several man-made materials (brick, concrete, ceramic, cinder block)
- safety goggles

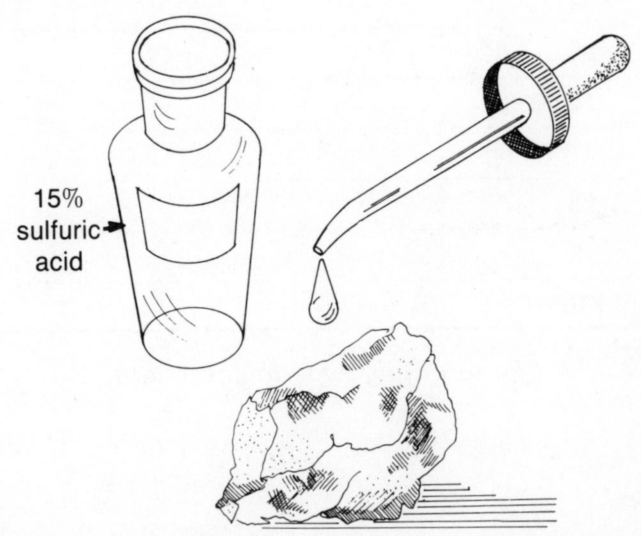

Fig. 4-6. *Using an eye dropper, place one drop of diluted sulphuric acid on each rock specimen. If it bubbles, calcium is present.*

Procedure

This experiment must be done wearing rubber gloves, safety goggles, and with an adult present. Place one drop of sulfuric acid on a test specimen (see Fig. 4-6). Observe whether or not bubbling occurs. After about 10 seconds of observation, place the specimen in the water in the bucket. The water will neutralize the acid. Record the results in a chart such as the one shown in Fig. 4-7. Proceed one specimen at a time and record in the chart. When the tests are completed, calculate the percentage correctly hypothesized. Were you able to determine calcium carbonate's presence by observation alone?

Going Further

Test other materials. For example, shells, clipped finger nails, chicken bones.

Specimen	Hypothesis	Calcium Present or Not
#1		
#2		
#3		
#4		
#5		
#6		
#7		
#8		
#9		
#10		

Fig. 4-7. *Chart for recording experimental data.*

5 Fossils

Fossils are evidence of things that once lived. There are three types of fossils. When an organism is completely encapsulated (enclosed) and preserved, it becomes a fossil. The object itself is the fossil. If a footprint or soft material such as a leaf imprints itself in mud, a fossil remains when the mud hardens into sandstone. This type of fossil is called an imprint. Calcium or some other minerals might "fill in" the cells of an object and harden, producing petrified stone. Organic material is replaced by minerals, such as petrified wood, where there is no cellulose (glucose that makes up the main part of cell walls) left.

Fossils do not occur in greater abundance because of decomposers. Most living things that die are eaten or decompose. Only items in very specific situations will become fossils.

PROJECT 1
Shoe Box Archaeology

Overview

Often archaeologists use the depth of a fossil material to help determine age, at least at the same site. For example, several towns might have been built on the same site after one "died out." Therefore, as an archaeologist digs down, the most recent items would be near the top, and the oldest items would be farther down. As items are being excavated, or dug out, they should be drawn in position in relationship to "north" and to all other items found. Archaeologists try to identify items and what they were once used for. Researching man's past helps us to better understand the present.

Materials

- shoe box
- ruler
- sand
- toy soldier or other action figure
- string
- plaster
- fish or chicken bones
- probes
- coins
- shark's teeth
- brushes
- dated newspapers
- resource books
- marking pens
- leaves
- wax paper

Procedure

Build a shoe box to be used as an excavation site by mixing two parts plaster to one part sand, and moisten to produce an even wetness. Using only plaster makes it too hard to dig into and too much sand makes the mixture too crumbly.

Put waxed paper into the bottom and sides of a shoe box. Place several objects in the box and cover with the sand-plaster mix. Fill the box half-way full. Then add more objects on top of the plaster, and cover them with more mix. Finally, as you pour in the rest of the mix, embed any remaining small objects.

Let the entire box set and harden. Mark an arrow on one end to indicate North. Now you are ready to "dig."

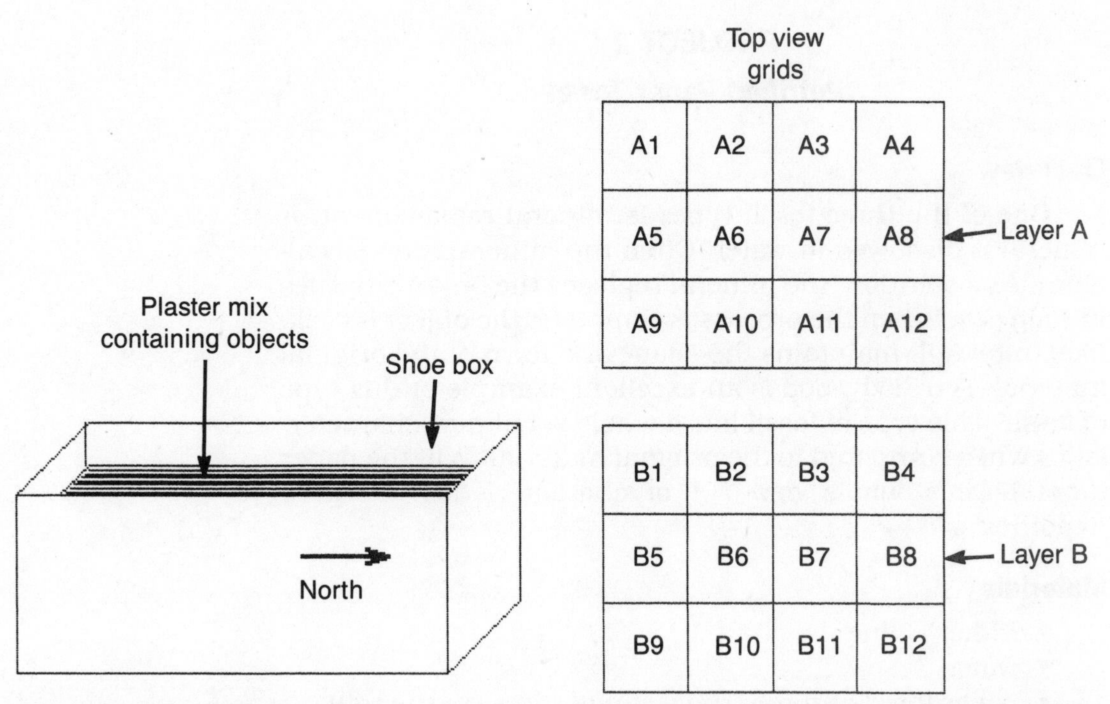

Fig. 5-1. *The concept of an archaeological dig can be demonstrated by constructing a simulated excavation area.*

Set up grids using string, and mark them as shown in Fig. 5-1. Probe or excavate gently in one section. When an object is located, go slower. Make notes when and where the object was struck. Remove as much material as possible. Clean specimens further with a brush. Hypothesize what it might be, and verify with a resource book. Continue searching using these steps.

Going Further

1. Set up a dig in an old dump (not used for 10 years or more).
2. What features were used to identify specimens?
3. Can relative age of objects be determined?
4. Can past features be determined?

PROJECT 2
Petrified Paper Towel

Overview

One of the three fossil types is mineral replacement. A mineral is dissolved in water. Often the mineral may be calcium. As a solution, the mineral replaces the organic tissues in each cell. When the process is complete, the object is rock-like, but it still maintains the shape and form of the original material. Petrified wood is an excellent example of this type of fossil. This type of fossil has a whole set of new characteristics when compared to the original material. Will the paper towel have a whole new set of characteristics? Form a hypothesis.

Materials

- Elmer's glue
- water
- paper towel and roll (the last end piece on the roll)
- mixing pan (about 8″ by 14″, or longer than the paper towel roll)

Procedure

In an 8″ by 14″ pan, mix a 2 to 1 portion of water to Elmer's glue (twice as much water as glue). The quantity should fill the pan to at least one-half inch depth. Start with four ounces of water and two ounces of glue. Use more as needed.

Roll the paper towel and its roll in the solution. Be sure all surfaces are moistened. Stand it on its end and let it dry. Follow the same procedure four more times, perhaps once in the morning and once again in the evening for two days.

Let the paper towel dry completely for at least one week. Test for new characteristics. Is it hard? Does it absorb liquids? Will it burn? Will it decompose as a paper towel does? Has it taken on new characteristics as you hypothesized?

Going Further

Can you improve on the mineral solution? Can you improve the test material? Try cardboard instead of a paper towel.

PROJECT 3
Print Evidence

Overview

Fossil imprints show evidence of past occurrences in nature. An imprint is produced by an object being pushed or pressed into a softer material. An animal stepping into clay, tar, or mud which later hardens, leaves an imprint. This is evidence of the animal's existence at that place and time. Can these imprints be identified? Would observers of our time be able to identify familiar objects imprinted in clay? Hypothesize how many objects your friends can identify from their imprints.

Materials

- 10 objects to be imprinted (examples: clothespin, pencil, paper clip, shell, golf ball)
- clay for embedding objects (the actual amount depends on the size of the objects you are imprinting)
- 10 paper plates
- 10 index cards
- 25 answer sheets
- 25 (or more) test subjects (people)

Procedure

Place a one-half-inch layer of clay in 10 paper plates. Be sure the clay has been kneaded into softness. Using each object one at a time, press the distinctive portion of the object into the clay. Prepare a small card by folding it in half. Number the index cards from one to ten. Place one numbered card next to each clay imprint. Prepare answer sheets in advance. Provide for the observer's name, age, date, sex, and answer spaces for the ten unknown prints (see Fig. 5-2). Allow many different people to test their skill. The greater the number, the more reliable your results. Log all answer sheets. Were older or younger people more accurate? Were there any differences between males and females? Were there objects hypothesized to create problems? Reach a conclusion about your hypothesis.

Name ______________________	Date ______________________
Male or female ______________	Age ______________________
Identify the imbedded objects:	
1. ______________________	6. ______________________
2. ______________________	7. ______________________
3. ______________________	8. ______________________
4. ______________________	9. ______________________
5. ______________________	10. ______________________

Fig. 5-2. *Print answer sheets for people to fill in as they test their skill at identifying fossil imprints in clay.*

Going Further

1. Set up a similar experiment to favor young children by the selected test material.
2. Set up a similar experiment to favor males or females by the selected test material.

PROJECT 4
Cryogenic Roses

Overview

Cryogenics is the study of the effects of low temperatures on objects and processes. Russian scientists discovered a wooly mammoth (an extinct elephant) frozen in the Siberian ice. They thawed it, cooked a piece, and ate it. It remained eatable.

Hypothesize that all structures frozen in ice will be preserved well.

Materials

- five rose buds just beginning to open
- four plastic margarine bowls (one-pound tubs)
- freezer
- water

Procedure

Fill four plastic margarine bowls with equal amounts of water. Pick five relatively equal rose buds that are just beginning to open. Note their fragrance if any is present. Submerge a rose bud in each bowl and place them in a freezer. Maintain the fifth rose at room temperature as a control. Observe daily.

At the end of one week, remove one bowl and let the imbedded bud and ice thaw at room temperature. Observe the bud's color, overall appearance, and texture (feel it). Does it have any fragrance? Record your observations in a chart such as the one shown in Fig. 5-3.

Rose Bud	Observations (smell, look, feel)
Frozen One Week	
Frozen Two Weeks	
Frozen Three Weeks	
Frozen Four Weeks	

Fig. 5-3. *Chart to record observation for the project* Cryogenic Roses.

A week later, take a second bowl from the freezer. Thaw, observe, and record your observations. The thawing time should be identical to the last one.

Each week, remove another frozen bud and evaluate until all have been thawed. Did the buds remain intact no matter how long they were frozen? Conclude whether your hypothesis was correct.

Going Further

1. Simulating the Russian wooly mammoth ice-casting find, repeat the above experiment, but use pieces of hot dog instead of rose buds.
2. Make ice castings using leaves, pine cones, pine needles, and other organic materials.
3. Research storage times for frozen poultry and meat. What factors affect shelf life? Temperature? Light?
4. Do frozen rose buds that have been thawed decompose at the same rate as those that have not been frozen? Does the length of time frozen make a difference?

PROJECT 5
Heat from the Past

Overview

The coal that is taken from the earth to use as fuel was formed long ago. Large ferns as tall as trees grew, fell, and decayed. More grew and fell. Over long periods of time, the growth and death of massive amounts of vegetation have continued. This material becomes compacted and compressed. As it is forced deeper into the earth, it encounters heat and pressure. These conditions cause it to form coal. Oil and gas are also formed in those areas. Is "soft" coal really softer than "hard" coal? Hypothesize which will be harder, and which will have a greater specific gravity.

Materials

- one piece of "soft" coal
- one piece of "hard" coal
- measuring cup with spout for pouring
- bowl
- Mohs scale of hardness test kit (See Resource List for suggested suppliers.)
- scale

Procedure

Use a Mohs scale of hardness test kit to test each specimen and record the results.

To measure for specific gravity, weigh a piece of coal on a scale. Record its weight. Weigh a dry bowl that will be used to catch displaced, overflowing water. Fill a measuring cup to overflowing with water. When the water stops flowing, place the dry catch bowl in position under the measuring cup's spout. Gently place the coal in the water-filled cup, causing some water to overflow out of the spout and into the catch cup. When it stops flowing, weigh the catch bowl with the water. Subtract the weight of the catch bowl to find the weight of the water. Record the weight of the water. Divide the weight of the water into the weight of the coal (make it accurate to two decimal places) to arrive at a figure for specific gravity. Follow the same procedure for both pieces of coal. Be sure to dry the catch bowl before each test.

Conclude from your results as to whether or not your hypothesis was correct.

Going Further

1. Using a globe, hypothesize an area that might be in the process of beginning to form coal.
2. Does "hard" coal give off more heat than an equal amount of "soft" coal?

6

Erosion

Erosion is the wearing away of a material. Particles of objects (rocks, statues, and so on) are loosened and transported away.

There are different types of erosion: water, ice, glaciers, wind and sand, waves, chemical, and temperature extremes. Ice can seep into cracks in rocks, freeze and expand. This expansion acts like a wedge, prying the rock apart. Rushing water in streams or the energy of wave motion can wear away shorelines. Water runoff from rain and melting ice or snow can erode soils. The greater the slope, the more erosion that will take place because of the velocity of the flowing water. When a rock is hot, it expands. When it cools, it contracts. These temperature extremes can cause small fragments of rocks to flake off. Acids produced by living organisms such as slime molds, lichens, mosses, and slugs break down rocks. Acid rain also causes weathering of objects. Strong winds can blow sand particles at high velocities against objects and cause their erosion.

Erosion is a slow process in most cases, but it constantly changes the features on the surface of the earth.

PROJECT 1
Up the Down Staircase

Overview

Abrasion can cause a wearing away by the scraping or rubbing of objects. People wear away the surface of the things they walk on, such as their shoes. Hypothesize that significant erosion takes place on steps that are frequently used by abrasion from walking.

Materials

- wooden set of steps that have an indentation in them where people walk
- micrometer (a device for measuring very small distances)

Procedure

Locate a wooden staircase that appears to have considerable erosion on the steps where people frequently walk (see Fig. 6-1). Using a measuring device, such as a micrometer or ruler), measure the thickness of a step at the edge by the railing where no one walks. Record this number. Measure the depth of an indentation in the board where people walk, probably in the center of the board. Compare the two figures. If treads are to be replaced on an old set of stairs, cut through

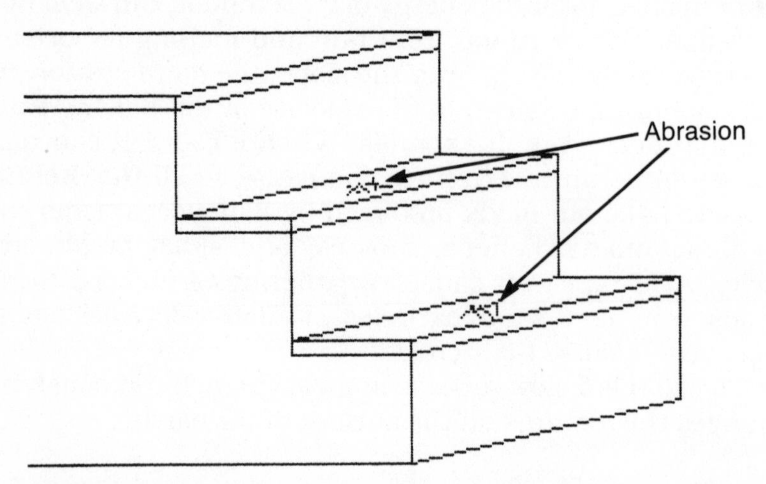

Fig. 6-1. *Measuring erosion due to abrasion from foot traffic.*

the worn portion of a step board to allow more precise measurement.

Conclude whether your hypothesis was correct.

Going Further

Devise a way to measure the difference between an unsheltered part and a sheltered part of an outdoor structure (patio, balcony, porch, deck). Where has weathering occurred to a greater extent?

PROJECT 2
Inky Dinky Spider

Overview

Rain gutters catch the runoff of rain from roofs, and funnel the water down a spout. The water pouring out of the end of the downspout can have considerable speed. This rapidly rushing flow can quickly wash away soil. Masonry materials, such as cement and brick, are often placed at the bottom of downspouts to bear the brunt of the raging water's force and disperse the rain over the ground in a less erosive manner. Hypothesize that by examining a brick that has been resting at the bottom of a downspout, you can estimate the length of time it has been in place.

Materials

- cement brick or block that has been at the bottom of a downspout for many years
- micrometer (a device for measuring very small distances)

Procedure

Locate a brick or block by a downspout (see Fig. 6-2). Look around your neighborhood. Be sure to get permission to be on someone's property. Estimate how long it has been there by the amount of erosion that has taken place on the brick at the point where most of the water hit. You can use a micrometer to make and record accurate measurements of the depth of the eroded area compared to other thickness areas of the brick. Conclude whether your hypothesis was correct by asking the homeowner how long the brick was in place.

You might want to locate other downspouts that have the same size and type of brick under them. From the information you gathered about the first brick (length of time it was exposed and the depth of the erosion), hypothesize how long other bricks have been in place.

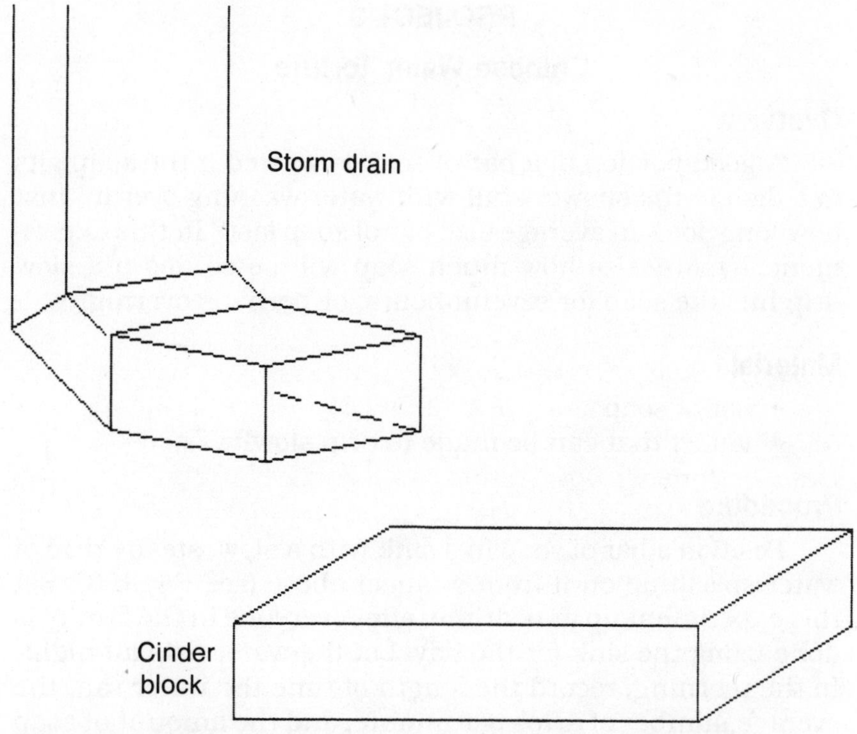

Fig. 6-2. *Measuring erosion caused by rainwater runoff.*

Going Further

1. Perform a Mohs test for hardness. A test kit will be needed.
2. Test different types of bricks that are for sale for the purpose of dispersing downspout water flow. Compare the different types by price and hardness (do a Mohs test for hardness). Determine which is the better buy. Would metal be better?
3. How can you deal with erosion around your home or school? Check rain downspouts. Roofs without rain gutters create erosion. Note the size of the soil particles being moved. The size of the particles depends on the velocity of flow.

PROJECT 3

Chinese Water Torture

Overview

A good portion of a bar of soap is wasted if the soap sits in a dish in the shower stall with water washing over it. Just how long does an average size bar of soap last? In this experiment, hypothesize how much soap will be eroded if a slow drip hits the soap for several hours, or perhaps overnight.

Materials

- bar of soap
- faucet that can be made to drip slowly

Procedure

Position a bar of soap in a sink with a slow, steady drip of water splashing on it from a faucet above (see Fig. 6-3). Set this experiment up at bedtime, after everyone in the family is done using the sink for the day. Let the water drip all night. In the morning, record the length of time the water ran, the average number of drips per minute, and the amount of soap that was washed away. You might want to weigh the soap on an accurate scale before and after the experiment to determine how much mass was eroded. Conclude whether your hypothesis was correct.

Going Further

1. Examine different soaps for hardness. Do some soaps last longer in the shower than others? How do these compare in price? Which brands are the better deals regarding price and length of time they are useful?
2. Larger rain drops, which occur in cooler weather, cause more erosion. Measure rain craters in fine powder.
3. Make a change in velocity due to height by using a shower head drip. Compare to a faucet.
4. Quantify by measuring the number of drips per minute or hour. Calculate the weight of 100 drips by collecting and weighing. Calculate the velocity of the drops due to height (32 feet per second each second or 32 feet per second squared).

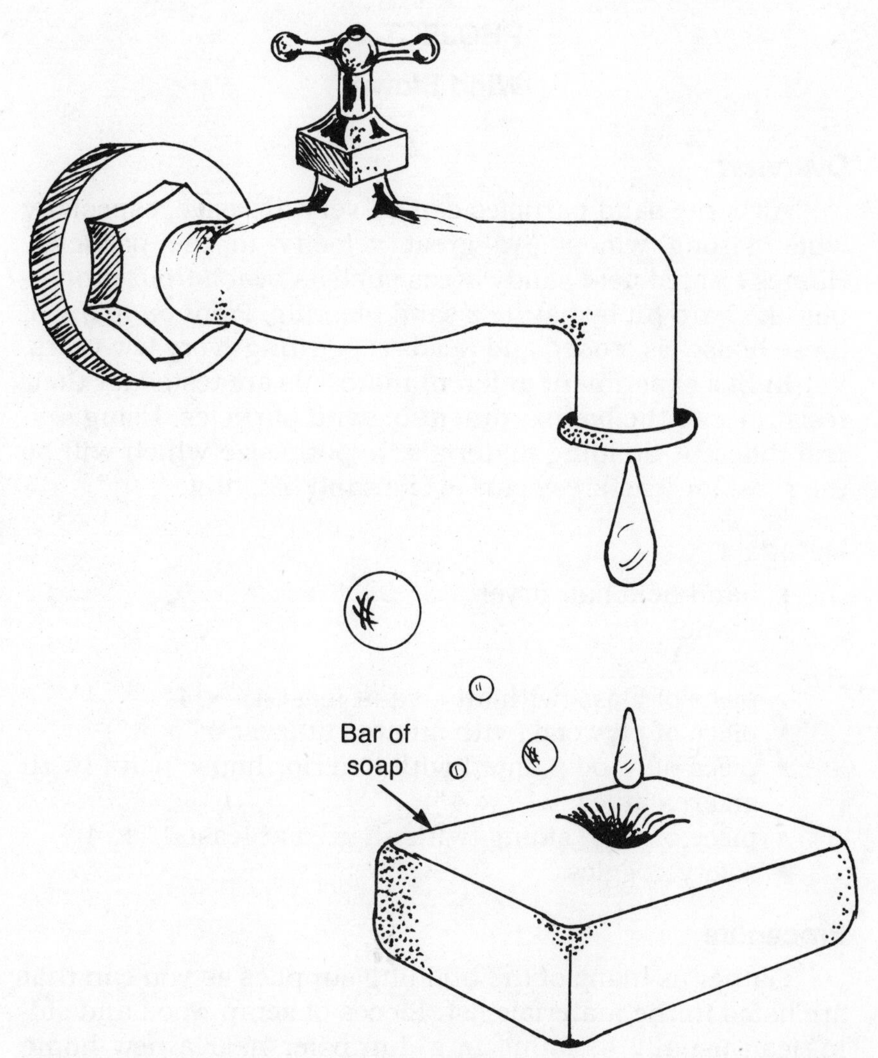

Fig. 6-3. *Measuring the effect of dripping erosion.*

PROJECT 4
Wind Blown

Overview

Airborne sand particles can be very abrasive, especially when strong winds give great velocity to the particles. Homes located near sandy areas such as beach front properties, are hard hit by nature's sand blasting. Paint on many of these houses is eroded and needs repainting every few years.

In this experiment different materials are tested for their resistance to the bombardment of sand particles. Using several different building materials, hypothesize which will be the most (or least) resistant to the sand blasting.

Materials

- hand-held hair dryer
- funnel
- sand
- piece of glass (with an area at least 4″ × 4″)
- piece of plywood (with an area at least 4″ × 4″)
- piece of wood painted with exterior house paint (with an area at least 4″ × 4″)
- piece of vinyl siding (with an area at least 4″ × 4″)
- safety goggles

Procedure

Gather as many of the building supplies as you can that are listed in the materials list. Pieces of scrap wood and siding can usually be found in a dumpster near a new home construction site. Ask a builder for permission to take a few scraps of different materials for your project. He might even be able to supply you with additional materials. If he is replacing old windows in a home, he may offer a pane of glass.

Aim a hair dryer at one of the building materials, such as a pane of glass. If the hair dryer has a heat switch, turn the heat off. You might want to have a friend hold the hair dryer. If the hair dryer has a speed switch, put it on the highest speed. Use extra care when handling materials, especially wood for splinters and glass for sharp edges.

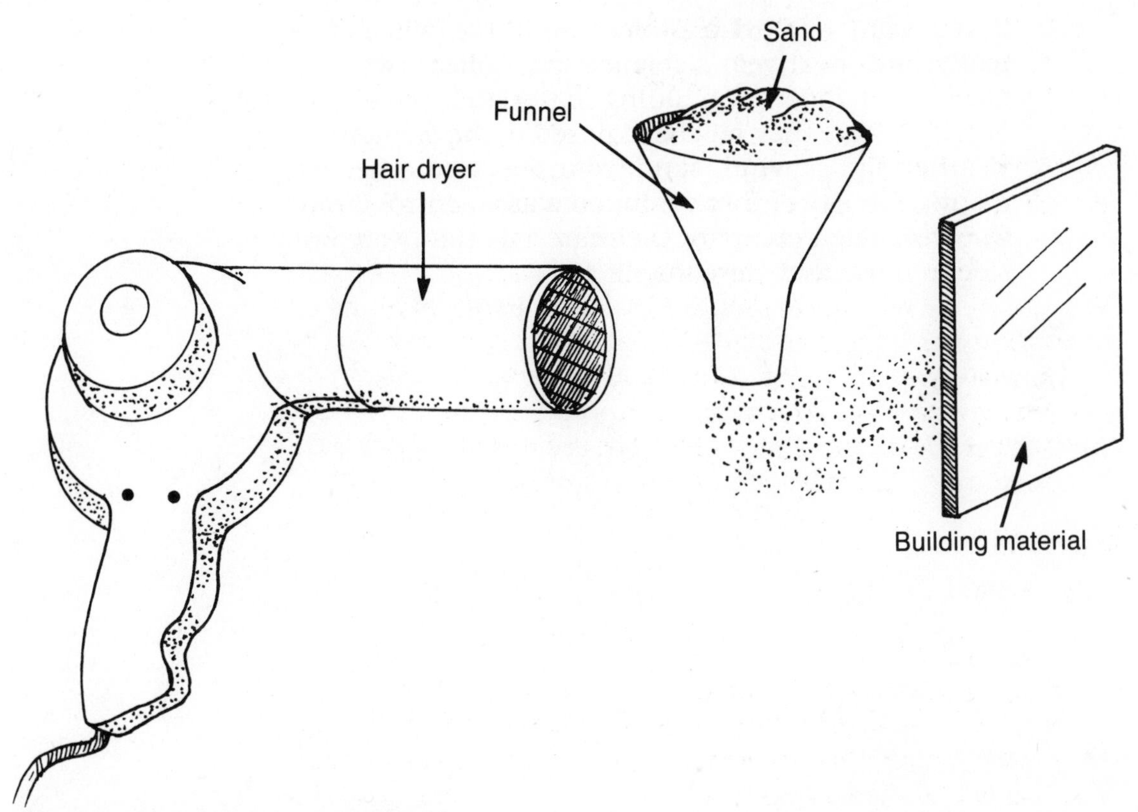

Fig. 6-4. *Measuring sand abrasion.*

Fill a funnel with sand. Place it about eight inches from the test material. Release the stream of sand coming out of the bottom of the funnel into the path of the moving air from the hair dryer (see Fig. 6-4). Fill the funnel again and repeat the process. Continue this procedure many times.

Examine the materials for pit marks or other signs of abrasion. Record your observations. Repeat this procedure for each different type of material you have available. Gather your results and conclude whether your hypothesis was correct.

Going Further

1. Do this project using different speeds on the hair dryer.
2. Do this project using different size sand particles—fine and coarse particles.

3. If you want to start a project now that will not be ready until next year's science fair, collect two samples of each different building material. Place one set outside where they will be exposed to the elements of weathering . . . wind, sand, rain, ice, and other elements. Place the other set indoors as a control group. One year later, compare the materials that were outside to those that were inside.

PROJECT 5
Throw in the Towel

Overview

Are sand particles transported by the wind to various heights? Using a towel, we will trap sand particles at different heights above the ground. Hypothesize that more sand will be trapped nearer the ground than up higher.

Materials

- five foot long towel
- two 2 × 4 lumber boards, six or seven feet long (or two poles that can be used to support the towel and hold it up in the air, such as volley ball net poles)
- sandy area (desert, beach, or area where there is no cover crop)
- two stakes

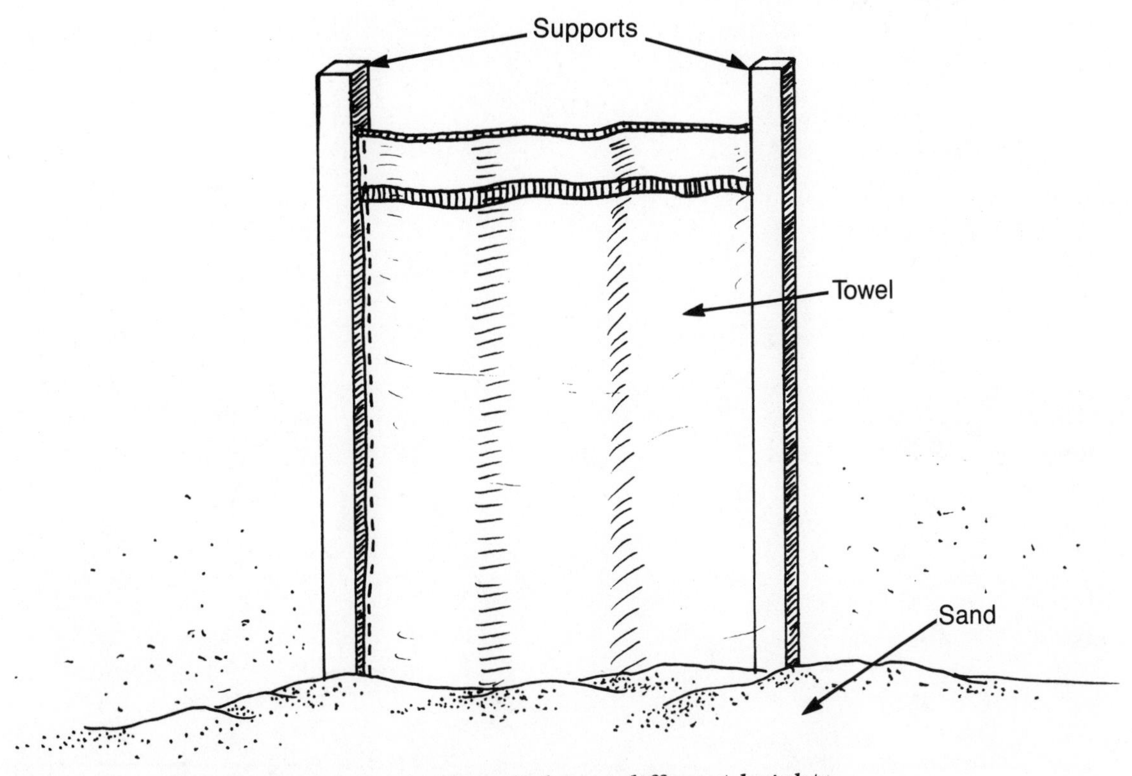

Fig. 6-5. *Collecting airborne sand particles at different heights.*

Procedure

Locate a sandy area. Using two support poles or boards, hang a five foot long towel lengthwise so that it starts at the ground and rises up five feet. Use stakes or some other method to firmly hold the bottom of the towel and keep it from swaying in the wind (see Fig. 6-5).

Wet the towel and keep it moist. The moisture should help trap and retain sand particles. After a period of time, take the towel down and evaluate the amount of sand that has accumulated near the bottom, the middle, and the top of the towel. Was your hypothesis correct or incorrect?

Going Further

1. Using a microscope, determine if there is any difference in the size of the particles found in the top of the towel as compared to those at the bottom.
2. People use windbreaks to stop wind erosion. Examine patterns created by wind erosion by placing a cinder block on sandy ground.

PROJECT 6

Automatic Sand Castles

Overview

Winds can transport sand particles over great distances. This movement of sand can dramatically change the features of an area. The sand particles carried by strong winds drop out of the wind and fall to the ground when the wind velocity slows down. Fences are sometimes erected to slow down the wind and let sand particles fall. Snow fences are placed along roadways so that snow will drift around the fences and not on cleared roads. The name snow fence is also used to describe windbreaker fences placed along beaches. These fences build sand dunes and maintain the sand that is already there (see Fig. 6-6). The infamous dust bowls of the 1930's in the Plains states were caused by drought and intensive farming. High velocity winds wildly blew top soil from one area to another. The dropping of sand or soil particles in an area is called deposition.

Fig. 6-6. *Snow fences, or windbreaker fences, are constructed along beaches to collect airborne sand particles.*

The spacing between the slats in a snow fence are significant. Hypothesize that there is an optimum slat separation distance for sand deposition.

Materials

- wooden box frame (about 2 × 4 or 5 feet, and 1 or 2 inches deep)
- beach sand or playground sand
- fan (preferably a three-speed fan)
- several dozen popsicle sticks
- ruler or yardstick
- glue

Procedure

Place a ruler on a flat surface, such as a tabletop. Lay popsicle sticks flat on the table in a row, spacing them at two-inch distances from each other (see Fig. 6-7). Glue several sticks lengthwise across them at the bottom to hold them all together. Refer to this group of sticks as set #1.

Repeat this procedure of building a popsicle fence, but build this second set with only one-half-inch spacing between the sticks. This will be called set #2.

Fill a 2 × 5 foot shallow box with sand. Make the sand fairly level. Bury the bottom of the first set of sticks into the

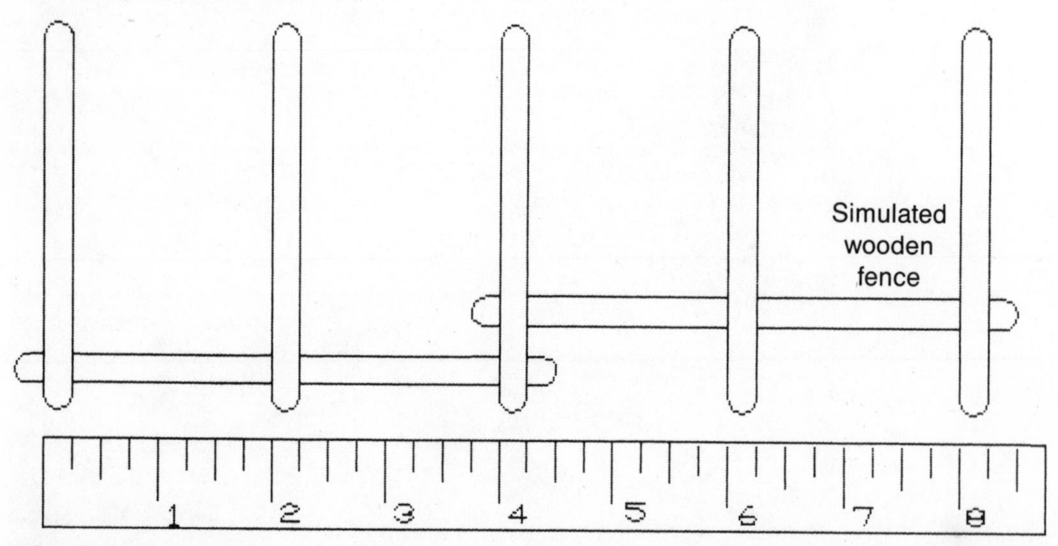

Fig. 6-7. *Construct a snow fence using popsicle sticks to determine optimum spacing between slats.*

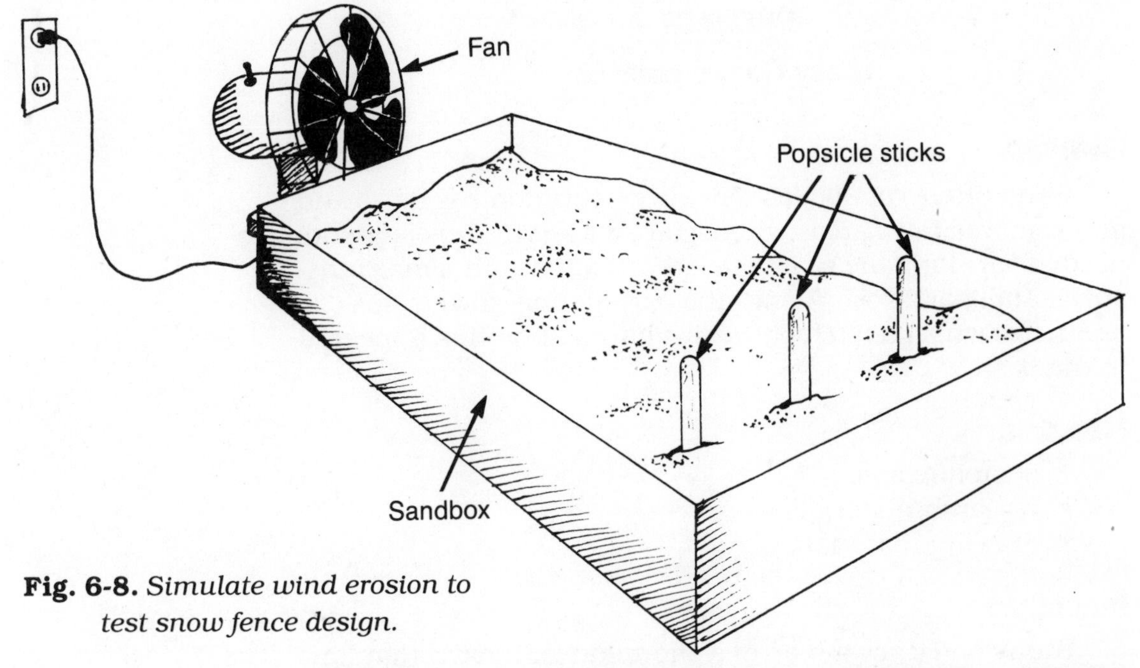

Fig. 6-8. *Simulate wind erosion to test snow fence design.*

sand near one end of the box. Place a fan at the other end (see Fig. 6-8). Turn the fan on for a short period of time. Observe, measure, and record any buildup of sand "dunes" around the popsicle fence.

Remove the first fence, smooth out the sand in the box, and bury the second fence. Turn on the fan. Compare the results of sand buildup between set #1 and set #2, and conclude whether your hypothesis was correct.

Going Further

1. What happens if the fence is a solid wall with no spacing between the slats? Hypothesize that the wind will simply blow over the fence without depositing any sand.

2. Sift sand into several different size granules and perform the above experiment using different size particles.

3. Try the experiment using different speeds on the fan, representing different wind velocities.

4. Could other shapes or patterns of snow fence slats trap sand better? Diagonal? Two rows, one behind the other?

PROJECT 7
Easy Come, Easy Go

Overview

Sand often gets deposited where you don't want it and gets removed from areas where you do want it. This is particularly true along the seashore, where water and wind transport sand particles. Along the coast, one town's loss is another town's gain. Hypothesize how a shoreline appeared in the past.

Materials

- shoreline area
- research materials
- drawing materials

Procedure

Study and draw a map of a shoreline for a town that borders the sea. Can you see any clues that might reveal a trend of how sand is eroding or being deposited? Have jetties been erected in the near past (see Fig. 6-9)? If so, you can hypothesize how the shoreline used to look. Draw how you think it

Fig. 6-9. *Jetties built along a shoreline alter the shape of the shoreline by affecting sand deposition.*

used to look. Research the area at the library or interview some long-time residents. Does it in fact look like it did years ago? Conclude whether your hypothesis was correct.

Going Further

Hypothesize that the depth of the water in front of a breakwater will be more, or less, than behind it in the bay. Use a sinker on a fishing line to measure depth.

PROJECT 8
Up on a Pedestal

Overview

Just as wind can transport sand particles, so can water. Snow fences or windbreaker fences are used to slow down the velocity of the wind to a point where sand particles drop out and accumulate. Hypothesize that by building a barrier in the path of moving water, the velocity of the water can be slowed down and sand particles caused to drop out and build up.

Materials

- access to a seashore area or a shallow stream
- two cinder blocks

Procedure

Place several cinder blocks on the beach or in a shallow stream at varying distances from the high-water mark. Put each block far enough away from the other to avoid interference. Put them in place at low tide and observe them at high tide (see Fig. 6-10). Hypothesize that, as the water moves around the blocks, it will move faster, and where the water confronts the blocks it will slow down. Do you think artificial barriers could be constructed to get sand to accumulate where people want it, or do you think the rushing water around the blocks will erode sand away, leaving the blocks up on pedestals of sand? Perform the experiment and use the results to conclude if your hypothesis was correct or incorrect.

Going Further

Using cinder blocks, divert a portion of the water flow in a small stream. Does this increase the water velocity and cause erosion?

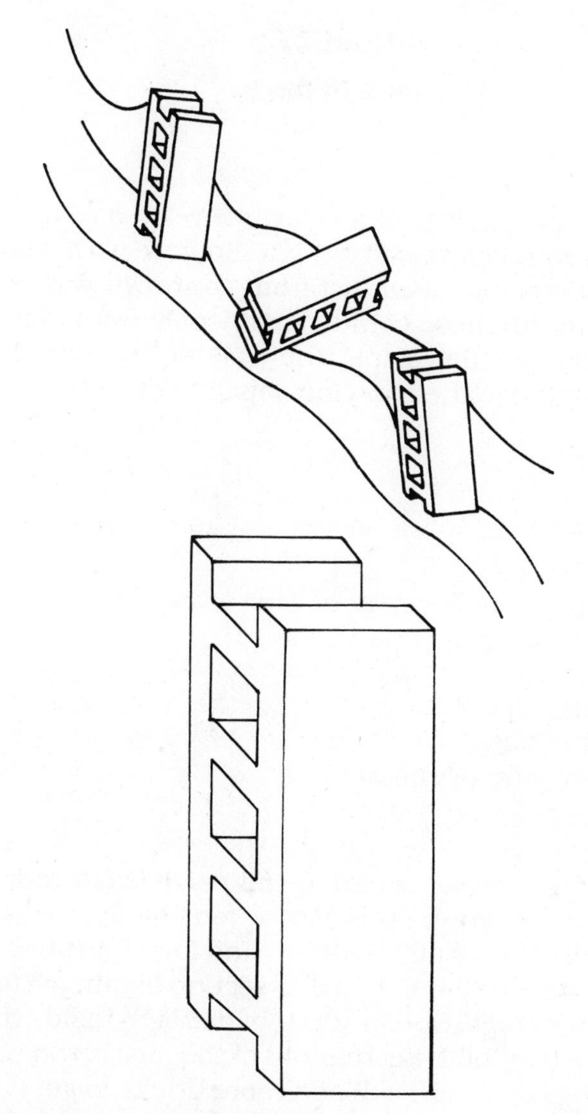

Fig. 6-10. *Use cinderblocks to determine their effect on sand particles carried by water.*

PROJECT 9
Perfect Pitch

Overview

The carrying ability of a stream is related to how much volume it has and its velocity. A garden hose with a one-half-inch diameter could clean dirt from your driveway. A three-inch diameter fire hose could clean people from your driveway! The steeper the slope, the greater the speed of the water. Hypothesize the carrying capacity of a stream due to its pitch.

Materials

- soil
- soil sieves
- five foot-long downspout pipe or rain gutter
- gallon water jug and water
- scale
- several bricks
- several bags
- square yard of cheesecloth

Procedure

Using a soil sieve, separate soil particles or stones into three or four separate sizes. Make four or five bags full of equal amounts of small, medium, and large particles. Using a five-foot-long downspout pipe or section of rain gutter, thoroughly wet the pipe. Spread a bag of material inside the pipe, lining the bottom of it. Set one end of the pipe up on bricks as shown in Fig. 6-11. You will add more bricks to get a steeper slope. Clear off the landing area at the bottom of the pipe. Place several folds of cheesecloth at the bottom to trap sand particles. Pour one gallon of water down the chute. Using a sand sieve, separate the particles that the water carried out of the pipe. Measure how much each size came out, perhaps by using a scale.

Completely clean out the pipe. Do this experiment again at different slopes. The pipe should be wet, otherwise your first run would be in a dry pipe and the others in a wet pipe, which might affect results. Use new cheesecloth.

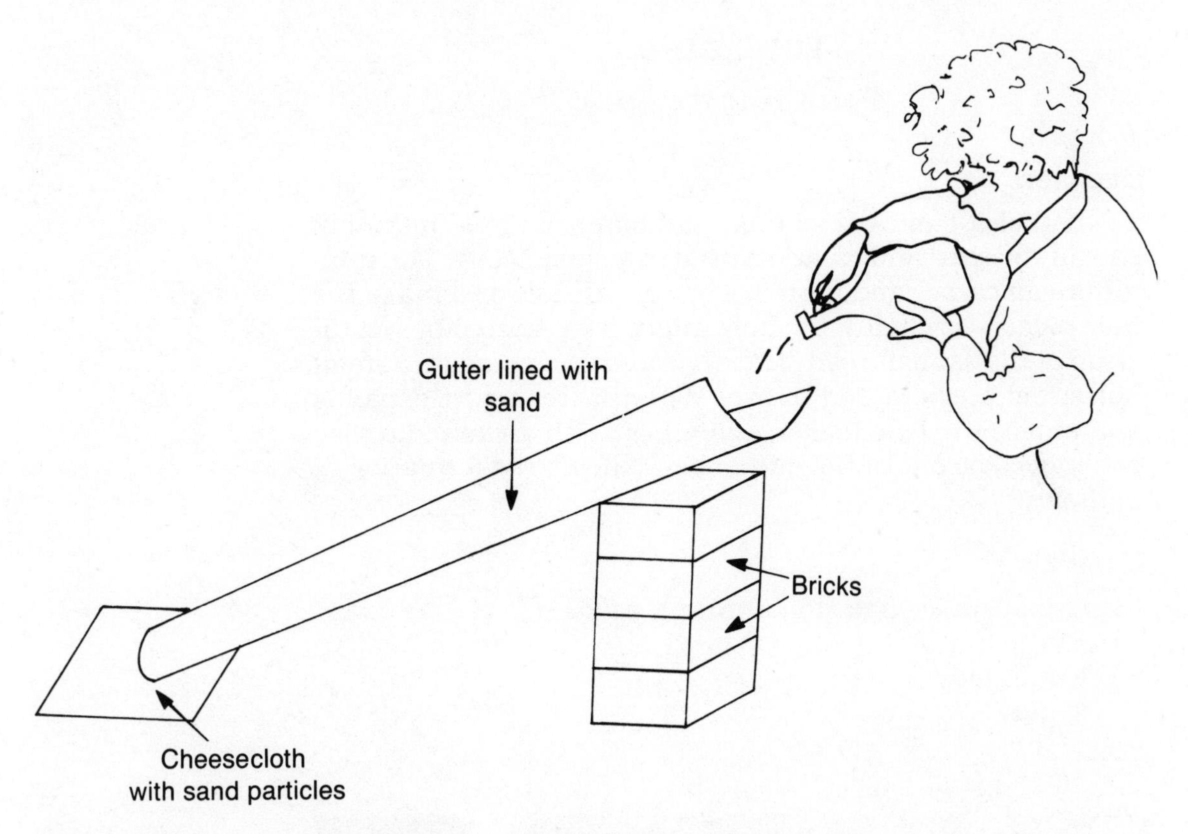

Fig. 6-11. *Determining the sand–carrying capacity of a stream.*

Going Further

Calculate the slope angle (the length and height of one end).

PROJECT 10
Potholes in the Road

Overview

Potholes can be a serious problem. Car tires hit small ruts in the road and cause rainwater to splash out. The rainwater can carry sand particles out of the hole and make the hole bigger. Originally, the hole might have started by weathering of the asphalt road surface. Water can creep into small holes and crevices and freeze. When it freezes, it expands and makes the hole bigger. Hypothesize that expanding ice can indeed be one of the influences causing the formation of potholes.

Materials

- piece of asphalt from a road surface
- water
- refrigerator
- glue
- ruler

Fig. 6-12. *Potholes in roads might be created by water expand- as it freezes in cracks in the asphalt road surface.*

Procedure

Find a piece of broken asphalt from a pothole or from the side of a roadway. Break it in half completely, then glue it back together, but leave some cracks. Pour water down into the cracks. Place it in a freezer. Try thawing, watering, and refreezing the asphalt several times. Does the expanding ice widen the crack? Measure and record your results. Conclude whether or not your hypothesis was correct.

Going Further

1. Survey your area. Determine where potholes occur with greater frequency: asphalt, stone, or gravel roadways. Measure the quantity of vehicles traveling on each surface over a given period of time. Are some road surfaces more prone to forming potholes than others?

2. Develop a relationship of vehicles using the road to the potholes. Either the number of vehicles or the type of vehicles (cars, heavy trucks, etc.) might be used.

7

Solar Energy

Most of the earth's energy comes from the sun. The heat inside the earth that produces thermal energy is not related to the sun, however. Nuclear energy is also not sun related, and the movement of tides is caused by gravity, not the sun's energy. But most energy comes directly or indirectly from the sun. Fossil fuels such as coal, gas, and oil were produced by solar energy. Plants grew, died, and were compressed under many layers of soil and rock for long periods of time. Therefore, fossil fuels are renewable. The length of time required to produce these fuels, however, makes society think of them as nonrenewable resources.

The renewable resource that comes immediately to mind is wood. Trees can grow and be harvested and grown again. Using energy from the sun in the photosynthetic process, trees and other plants grow. This trapped energy can be released as heat when the wood is burned. Peat and dung from animals, also used for burning, can be traced back to the sun's energy.

The sun heats the atmosphere, which causes air movement, such as that used for windmills. The warming effect on the surfaces of bodies of water causes evaporation, which permits precipitation. Precipitation makes hydroelectric energy possible. As water falls through generator mechanisms, electricity is produced.

Heat from the sun can also be stored in rocks and liquids and released at a later time.

Solar energy can be used directly to produce energy too. Photovoltaic cells can convert sunlight directly into electricity.

In this chapter, we present some project ideas for harnessing the power of the sun to help mankind on earth.

PROJECT 1
Solar Distiller

Overview

One fantastic property of water is evaporation. It can change its physical state from a liquid to a gas. A puddle, lake, stream, ocean, or any other body of water will allow some of its surface material to leave as a gas. When this occurs, any material other than H_2O (water) remains with the liquid. Therefore, if trapped and condensed (returned to a liquid state), the water is considered to be clean or distilled. The earth accomplishes this naturally by continuously producing precipitation in the form of rain, snow, hail, or fog. Can this process be improved? Can we build a passive solar distillation device to produce pure water from dirty water without having to use any fossil fuel energy to do it? Can clean distilled water be made faster by using a lens to concentrate sunlight? Hypothesize that we can make pure water passively by using the energy from the sun to evaporate unclean water.

Materials

- wooden dowel, about 3/8″ × 6″ (or a long pencil)
- two, two-liter plastic soda bottles
- two, three-liter plastic soda bottles
- three straws
- piece of plastic food wrap, about one square foot
- scissors
- seawater, saltwater, or soft drink
- liquid measuring cup, ounces or milliliters
- brass paper fasteners
- masking tape

Procedure

Cut a two-liter plastic soda bottle in half, but as you cut, carve out a hook on two opposite sides so that the bottle bottom can be hung from a pencil or wooden dowel as shown in Fig. 7-1. Cut the top off of a three-liter plastic soda bottle. Punch two holes on opposite sides of the three-liter bottle. The holes should be near the top. Fill the two-liter container with unclean water. Use salt added to water, seawater, or a

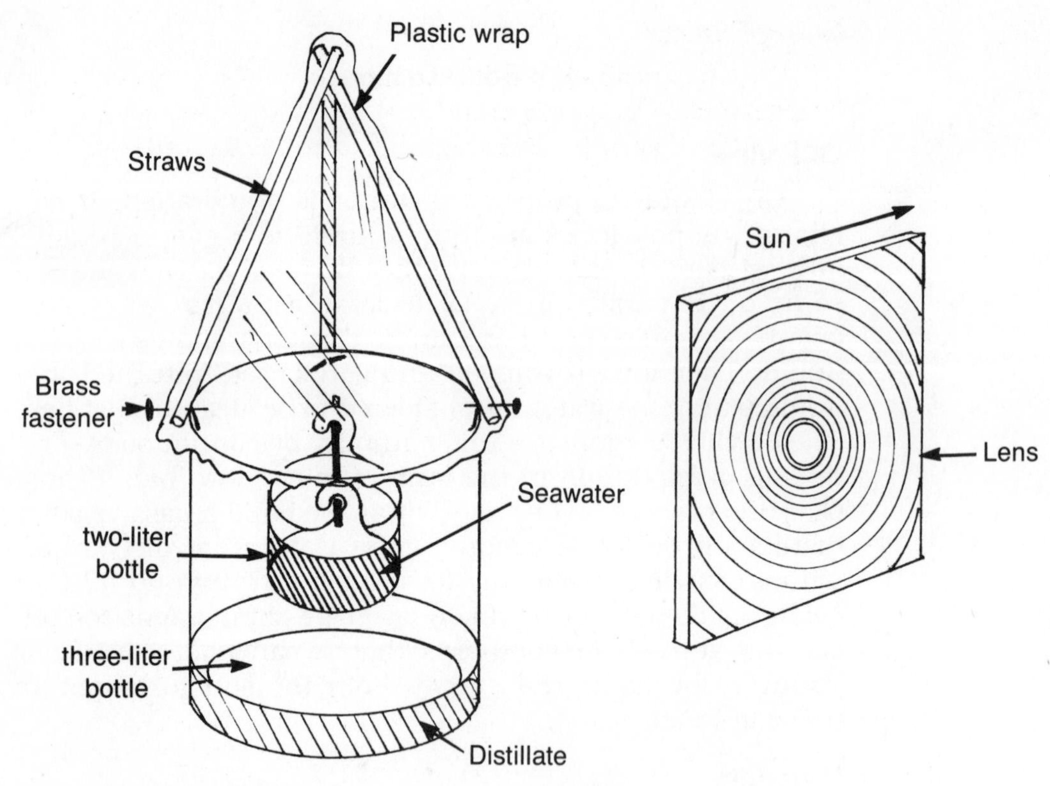

Fig. 7-1. *Produce clean water by solar distillation.*

soft drink. Place the two-liter container inside the three-liter container.

Insert a wooden dowel or pencil through the two holes in the three-liter container and lift up the two-liter container and hang it on the stick so it is suspended. Using several straws, masking tape, and a square of plastic food wrap, construct a tentlike top that fits over the three-liter container. The plastic wrap must extend down into the three-liter container. Use brass paper fasteners to hold the hood in place. The tent will act as a hood to trap any evaporating water vapor. Water vapor should condense inside of it and drip down the sides, collecting at the bottom of the three-liter container.

Place your solar still in a sunny location, but try to keep the top tent hood out of the sun. If the tent top is cool, it will condense water better. Conclude whether or not your hypothesis was correct.

Going Further

Construct the solar still and measure the amount of time it takes to distill a given amount of water. Perform the experiment again, with the same amount of sunlight and temperature, but this time place a Fresnel lens in the path of the sunlight to concentrate the sun's rays on the two-liter container of unclean water. Compare the two lengths of time. Was adding a lens worth it?

PROJECT 2
Keep Warm

Overview

One of the least expensive methods of supplemental home heating is passive solar. Passive solar refers to devices that collect energy from the sun and release heat to the home without the device itself requiring any additional energy to work. If heat from the sun can be stored during the day and slowly released during the cooler evening hours, it would be a very useful heat source. In this experiment, different materials will be tested to see which will radiate heat the slowest, thus making it a good heat source in a home. Hypothesize which materials will store heat the longest and test them.

Materials

- four, three-liter plastic soda bottles
- four long thermometers
- flat black paint and paint brush
- sunny window
- soil
- medium-sized stones
- water
- four, one-hole rubber stoppers that fit the neck of the soda bottles
- thick gloves

Procedure

Paint four three-liter plastic soda bottles black. Because color affects heat absorption, all must be the same color to absorb an equal amount of heat during the day. Fill one with soil from your backyard, one with medium-sized stones, and one with water. Leave the fourth filled only with air. Wear gloves and carefully (glass thermometers can break) push a thermometer through each rubber stopper so that the bulb end of the thermometer will be about in the middle of the bottle when the stopper is pushed into the top (see Fig. 7-2). If you do not have a rubber stopper, corks could be used, and a hole of the appropriate size drilled in them (adult supervision required when using a drill).

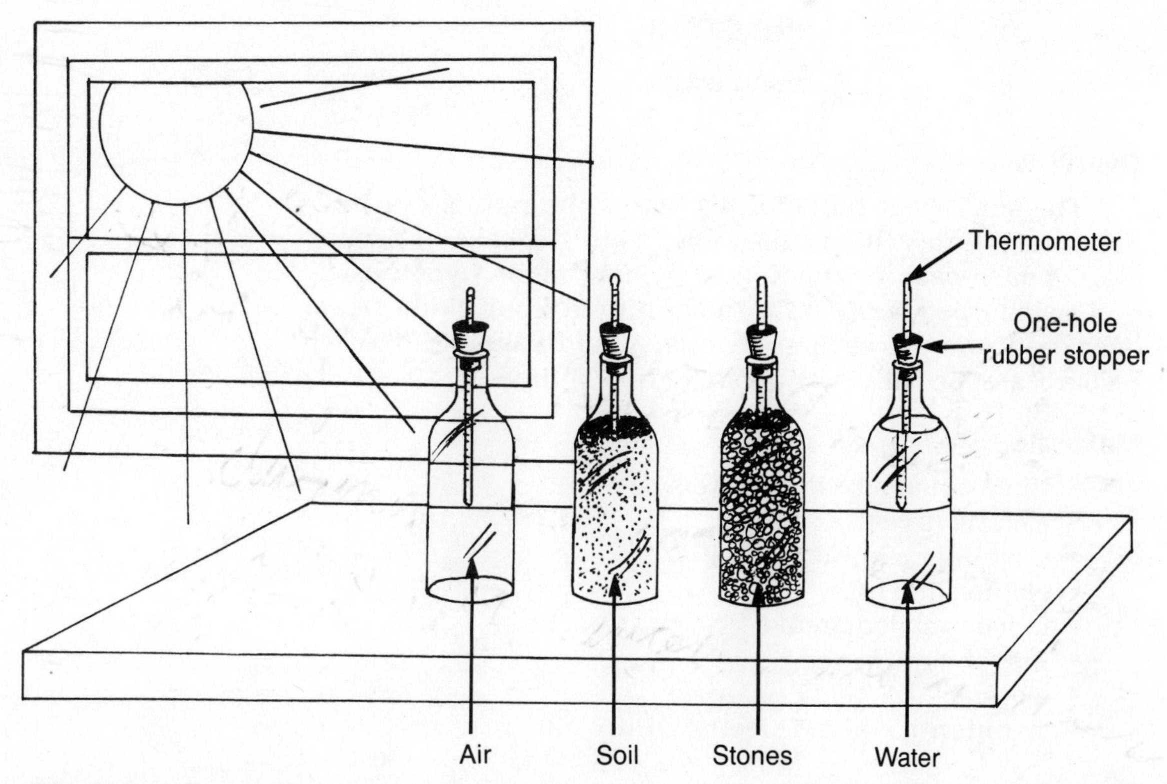

Fig. 7-2. *Determining which types of materials are best suited to store heat collected from the sun.*

Place the four containers in a sunny window. Record the temperatures of each bottle before sundown. Every hour after sunset, measure and record the temperature of each solar storage sample. Make a log to chart the temperatures over time. From your data, conclude whether your hypothesis was correct.

Going Further

1. Try different materials: wood, glass, paper, rock. Test solid materials by drilling a hole in them and inserting the thermometer, such as a brick.
2. Try stones in water and soil in water, comparing both to water alone. Water has a faster heating capacity but stones retain heat longer. Are combinations better than the individual materials alone?

PROJECT 3
Hang It Up

Overview

The sun's heat radiation increases the rate of evaporation. Many times this is desirable, as in the case of drying clothes after washing them. In this experiment, hypothesize that using free energy from the sun to dry our clothes can offer a substantial savings in money and valuable natural resources, especially over a long period of time.

Materials

- three equal-size bath towels
- electric dryer
- washing machine
- clothes line rope
- indoor shaded area
- indoor sunny area
- clock
- last month's home electric utility bill

Procedure

Thoroughly soak three bath towels of equal size. Place them in a washing machine and put it in the spin cycle. This will wring out excess water equally in all the towels.

Set up a clothes line in a shaded indoor area. Hang one of the towels on the line. Set up a clothes line in a sunny indoor area and hang a towel on it. Place the third towel in an electric dryer. Record the starting time (use a clock and write the times on the chart in Fig. 7-3). Periodically check on all three towels. Record the time that each is fully dry.

In the owner's manual, or on a plate on the back of the electric dryer, there is a listing for the number of watts consumed by the dryer. Look at last month's electric bill from a utility company. Calculate how much one kilowatt of electricity costs. This can be done by dividing the total dollar amount of the bill by the number of kilowatts used.

$$\text{cost per kilowatt} = \frac{\text{bill amount}}{\text{kilowatts used}}$$

Towel	Clock Time Start	Clock Time End	Drying Time
Shaded Area			
Sunny Area			
Dryer			

Fig. 7-3. *Record drying times for three identical towels placed in shade, sunlight, and in an electric dryer.*

Therefore, if an electric bill was $64.75 and 619 kwh (kilowatt hours) were used, then the cost per kilowatt was about 10 cents. Power companies have different rates for varying amounts of energy used, but this should give you a good approximation.

Armed with this information, determine how much money it cost to run the dryer to dry the towel. The formula to use is:

$$\text{cost} = \text{watts} \times .001 \times \text{kwhcost} \times \text{hours}$$

Watts is the number of watts used by the dryer, *.001* converts watts to kilowatts, *kwhcost* is the cost per kilowatt hour, which you just computed, and *hours* is the length of time in hours it took to run the dryer. If the dryer is rated at 1,200 watts, the power cost 10 cents per kilowatt hour, and the dryer had to run for a half hour (.5), then:

$$\text{cost} = 1200 \times .001 \times .10 \times .5$$
$$\text{cost} = 0.06 \text{ or } 6 \text{ cents}$$

Conclude from your data whether your hypothesis was correct.

Going Further

1. Add wind as a factor by placing the towels outdoors.
2. Design an experiment that allows even evaporation. For example, a towel hanging on a clothesline will dry at the top before the bottom because of gravity.

PROJECT 4

Good Mirror, Bad Mirror

Overview

The reflective properties of a mirror can be tested with aluminum foil by bouncing sunlight off of them and into containers of water. The water will warm up if it receives energy from the sun's rays. The warmest water would indicate the best reflector. Hypothesize whether or not the aluminum foil is as good at reflecting the sun's rays as the mirror.

Materials

- two, small blocks of wood
- two thermometers
- two, clear glass jars
- sunny window
- sheet of cardboard
- aluminum foil
- water
- mirror
- adhesive tape

Procedure

Cut out a cardboard sheet the same size as the mirror you will be using. Fasten the foil to the cardboard using adhesive tape on the back.

Fill two jars, such as mayonnaise jars, with water. Insert a thermometer in each. Place them in the shade near a sunny window. Keep them out of direct sunlight. Using a small block of wood or other means of support, place the mirror where it can reflect the sun's rays directly into one of the jars (see Fig. 7-4). **Do Not Look Into The Sun. Do Not Shine Sunlight Into Anyone's Eyes—Be Careful.**

Angle the foil reflector so that it focuses light on the other jar. Wait awhile for the water to heat. Check the setup occasionally to be sure the sun is still being reflected into the containers. Remember, the earth is moving and the sun does not stay in the same spot in the sky. Slight adjustments will be needed to the reflectors.

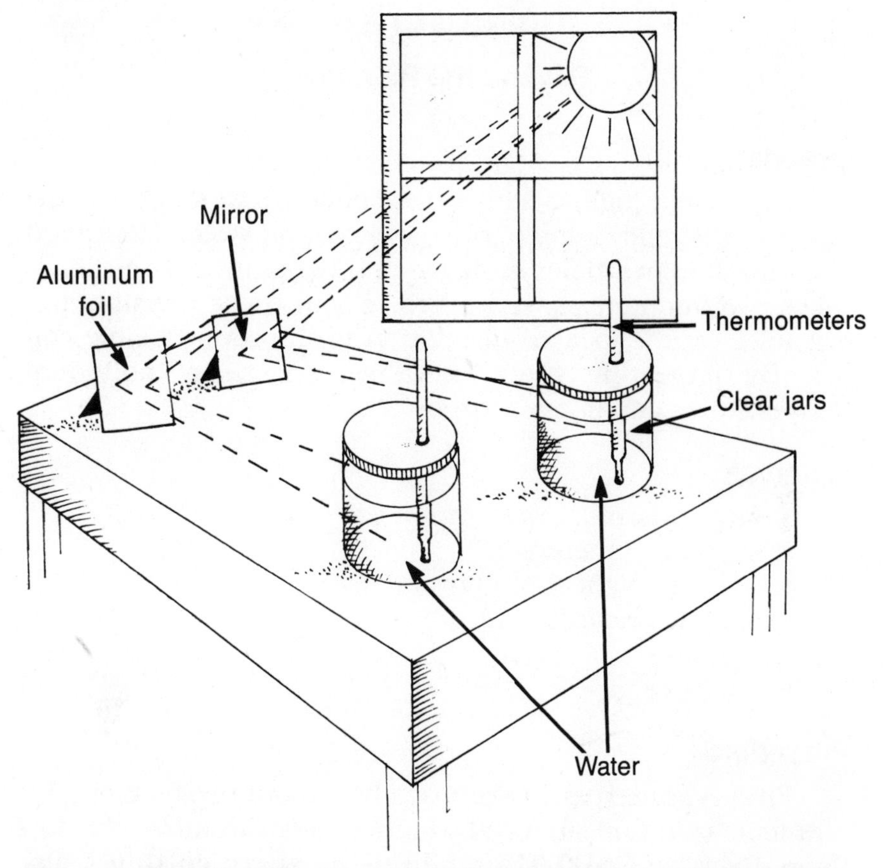

Fig. 7-4. *Test the reflective properties of different reflective materials. Keep the jars of water out of direct sunlight. Only reflected light must hit them.*

Read the temperatures on the thermometers and conclude whether your hypothesis was correct.

Going Further

Compare the reflective properties of other materials: wood, metal, clear glass, and any other materials you wish to try.

PROJECT 5
Beyond the Rainbow

Overview

The visible light spectrum is made up of seven colors: red, orange, yellow, green, blue, indigo, and violet. Below red is infrared. Above violet is ultraviolet. We cannot see either of these two frequencies. Hypothesize that these invisible frequencies can be located in relation to the seven visible colors. By projecting them on a screen, their individual temperatures can be measured.

Materials

- large prism
- six thermometers
- 2 × 2 foot sheet of plywood
- oak tag or cardborad
- markers
- adhesive tape

Procedure

Find or construct a cardboard box about two feet long by one foot wide and about six inches deep. Remove the top flaps. Exercise **CAUTION** when using sharp cutting tools. Cut a piece of oak tag paper about one foot high and about two and a half to three feet long. Curve the oak tag to fit inside the cardboard box, making an arc like a movie screen in an amphitheater. Use adhesive tape to secure it.

Place the box at one end of the plywood sheet, with its open side facing toward the spacious end of the board (see Fig. 7-5). Place a large prism at the spacious end. Position it so that when light hits it, the colors of the spectrum will be displayed on the screen. Depending on the size of the prism, you might have to experiment with the position of the prism and screen. The top side of the cardboard box will act as a hood to shield sunlight from shining onto the screen where the spectrum colors are being displayed. Only the prism itself should be in sunlight.

Put the prism in sunlight and mark off the locations of the spectrum colors using a pen or marker. The next frequency above violet is ultraviolet and below red is infrared.

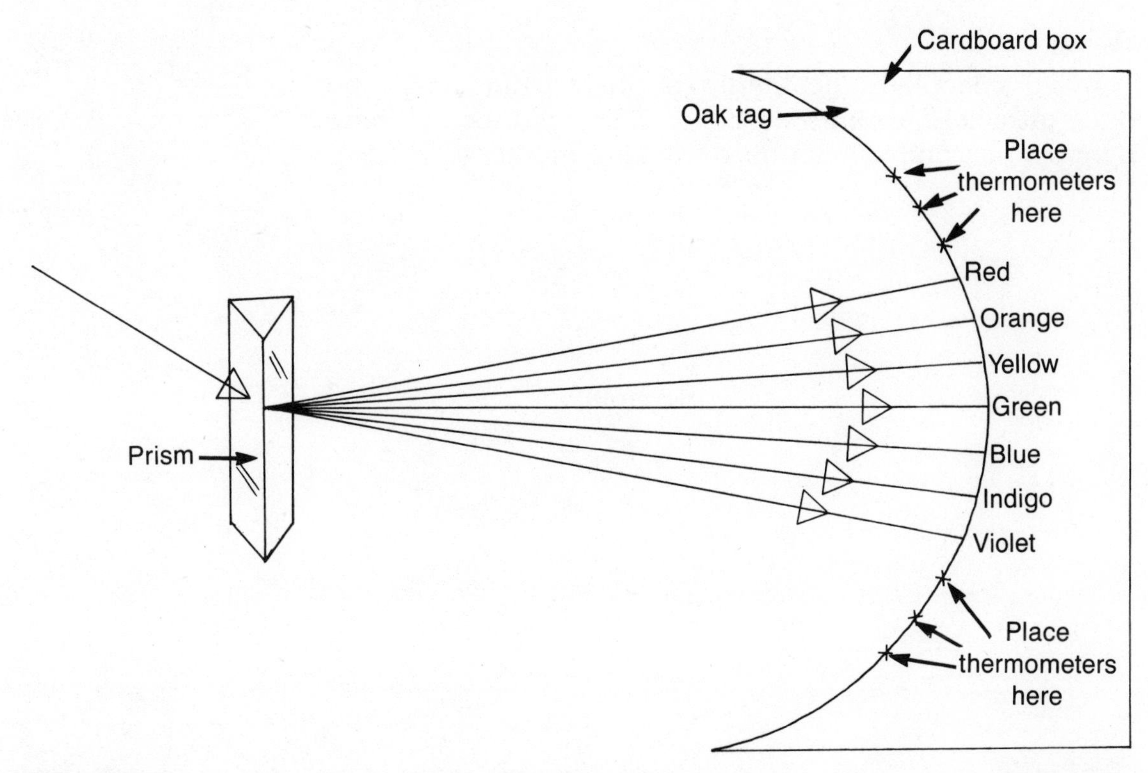

Fig. 7-5. *Locate where ultraviolet and infrared light appear on the spectrum by temperature sensing.*

Try to locate these by positioning three thermometers in each area. Leave a little space between each. You might need to adjust the spacing to locate the exact points. A thermometer, which is "on center" of the invisible frequencies, should read a higher temperature than neighboring ones. Because the earth is moving, you will only have a few minutes to line up your spectrum colors to their screen markings and measure temperatures in the invisible areas. You might want to devise a method of tracking the sun so that you will have a longer time to take temperature measurements. The whole project could be placed on a "lazy susan" or a mirror could be used and constantly adjusted to keep light focused on the prism.

Based on your observations, conclude whether your hypothesis was correct.

Going Further

Can a floodlight, heat lamp, or other artificial light produce ultraviolet or infrared light? These will be stationary light sources and will not need tracking methods.

PROJECT 6
Hot Colors

Overview

The seven colors in the visible spectrum are different frequencies of light. Is one hotter than the other? Can any differences in temperature among the colors be measured? Hypothesize that it can, and which color you believe is the hottest.

Materials

- large prism
- seven thermometers
- 2 × 2 foot sheet of plywood
- oak tag
- markers
- adhesive tape

Procedure

Construct the screen setup explained in the last project, *Beyond the Rainbow*. Place the structure in a sunny place so that the sun shines on the prism but not on the display screen. Using adhesive tape, secure each thermometer to the center point of each color. Measure and record the temperatures of each. Because the earth is moving, you will only have a minute or two to make all measurements before you must readjust because of the shifting position of the sun. Conclude whether your hypothesis was correct.

Going Further

Which color-filter paper would be best to put on windows in the winter to let the most heat in? Which color would be the best to make an awning with to keep a porch cool?

8 Weather

The air and its contents that surround the earth is constantly moving and changing. This daily change is called weather. The air, or atmosphere, that surrounds the earth consists of moving air (wind), gases, tiny particles, and water vapor (see Fig. 8-1).

Generally, the term weather is used when referring to daily local conditions, such as temperature, humidity, snow, rain, and high winds. The word climate refers to long-term weather averages of a larger area. For example, we might talk about the average rainfall each year for a given region. Climate changes directly affect the surface features of a given region, such as deserts and glacial ages.

Every living organism on the face of the earth is affected by weather. Weather determines how you will dress, what outdoor activities you plan (a summer barbecue for instance), when to take steps to protect your home (impending tornados or floods), and how much food farmers can grow for our consumption. Weather can help mankind by providing rain water to grow crops, or it can be devastating to property and life. Between 1959 and 1987, 2,801 people lost their lives due to lightning. People have been killed by talking on the telephone during a thunderstorm. In the state of New York, a farmer plowing his field sustained injuries when

Photo courtesy of National Oceanic and Atmospheric Administration.

Fig. 8-1. *A U.S. cloud cover as seen from an orbiting weather satellite.*

lightning struck him on his tractor. As the ambulance drove him to the hospital, it too was struck by a lightning bolt, causing the vehicle to crash, killing the farmer.

Flash floods caused by heavy rains kill more people than any other weather phenomena. Obviously, gaining knowledge and understanding the weather is not only fascinating but it is also essential to our lives. As methods of forecasting improve, more and better preparation for violent weather is possible.

Experiments in weather can deal with air movement, the weight and pressure of air, water vapor in the air (rain, snow, and dew point), and weather forecasting.

PROJECT 1
Observational Weather Forecasting

Overview

By carefully observing the conditions of the air and sky around us, we can predict the weather for our local area fairly accurately. This prediction should be accurate for at least the next several hours and possibly for up to 24 hours. Hypothesize that a short-term accurate weather forecast can be made simply by gathering and interpreting observational data. These predictions will be done without the aid of any weather measuring equipment.

Materials

- pad and pencil
- cloud pictorial identification chart
- research materials

Procedure

Gather information on observable weather factors from research materials, such as encyclopedias and books on meteorology. With this data, construct your own charts such as the one shown in Fig. 8-2. It gives a list of cloud names and the kind of weather they usually bring. Your research should enable you to make charts that you can reference as many observational factors as possible. For example, how should you interpret seeing a ring or halo around the sun or the moon (ice crystals in the upper atmosphere)? Is a particular type of cloud moving toward you or away from you? Does it feel like it is getting warmer or colder outside? If the daytime sky has no clouds in it, does that mean the approaching nightfall will be cool because there are no clouds to act as insulation, keeping the warm air from escaping?

Using the charts you have compiled, begin to make observations and forecasts. Keep a record each day for several weeks of your forecasts and the actual weather developments. Compare the accuracy of your forecasts to what the actual weather turned out to be. Conclude whether or not your hypothesis was correct using the data you collected.

Weather Forecasting by Cloud Type Table

Cloud Name	Forecast
Cumulus	Fair weather ahead If they are building, it may turn stormy
Cumulonimbus	Rain - be alert for strong wind gusts and possible lightning
Stratocumulus	There is a chance of light rain If it is cold, there is a chance of snow
Altocumulus	If these are accompanied by a halo around the moon or the sun, rain could be coming
Cirrocumulus	Expect a change in the weather
Cirrostratus	Expect a change to bad weather

Fig. 8-2. *Cloud names and the usual type of weather they bring.*

Going Further

1. How does the addition of a single measured factor increase the accuracy of predictions? For example, does knowing if the barometric pressure is going up, down, or holding steady help in addition to the observational data? How about the direction of the wind?

2. Calculate the percentage of accurate predictions for four hours, eight hours, and 24 hours over a period of time. As the duration of your prediction increases, does the accuracy of the prediction go down?

STOP

PROJECT 2

The Dews And Don'ts

Adult Supervision Required

Overview

When water vapor in the air condenses and forms droplets of water on the grass, cars, and other objects, it is called dew. Dew occurs when the air becomes saturated and cannot hold any more water vapor. Consider the air temperature and the amount of water vapor present. If the pressure and moisture content remains the same but the temperature gets colder, as it would during the evening hours, at what temperature would dew begin to form? This temperature is called the dew point.

The dew point is generally defined as the temperature to which moist air must be cooled for saturation to take place. Knowing the dew point can be important. This measurement can predict the formation of fog as well as dew. If the temperature is below freezing, then frost will form. This is called the frost point. When the dew point is high, there is a lot of moisture in the air. Comparing the difference between the present air temperature and the dew point can tell us about the relative humidity. When the dew point and the air temperature are far apart, the relative humidity is very low.

Hypothesize that we can measure the dew point by cooling a metal surface until water droplets form on it and then measuring that surface temperature.

Materials

- strip of metal, about one inch wide and four to six inches long
- two-liter plastic soda bottle
- utility knife
- water
- thermometer
- rubber bands
- several small wooden blocks
- warm or hot day

Procedure

Because warm air can hold more moisture than cold air, we suggest you perform this experiment on a warm or hot day. Also, it should be a calm day. The dew point is colder than the air temperature. The stronger the wind, the closer the surface temperature of an object will be to the air temperature. In this case, it is not likely to drop below the air temperature. Furthermore, the greater the wind velocity, the more the moist air above the surface of the object is mixed with drier air. This makes a lower relative humidity. The lower the relative humidity, the lower the dew point. You might have noticed that dew does not usually form on objects during windy nights.

Fill a two-liter plastic soda bottle with water. Place it in the freezer overnight. The next day, have an adult cut away the plastic bottle using a utility knife. Discard the bottle parts, leaving a giant ice cube. Lay the bottle-shaped ice cube on its side (see Fig. 8-3). Push several small blocks of wood against the sides to keep it from rolling. Using rubber bands, strap a piece of metal lengthwise along the top of the ice cube. Using another rubber band, strap a thermometer to the top of the metal strip. Be sure the metal bulb part of the thermometer makes good contact with the metal strip. You

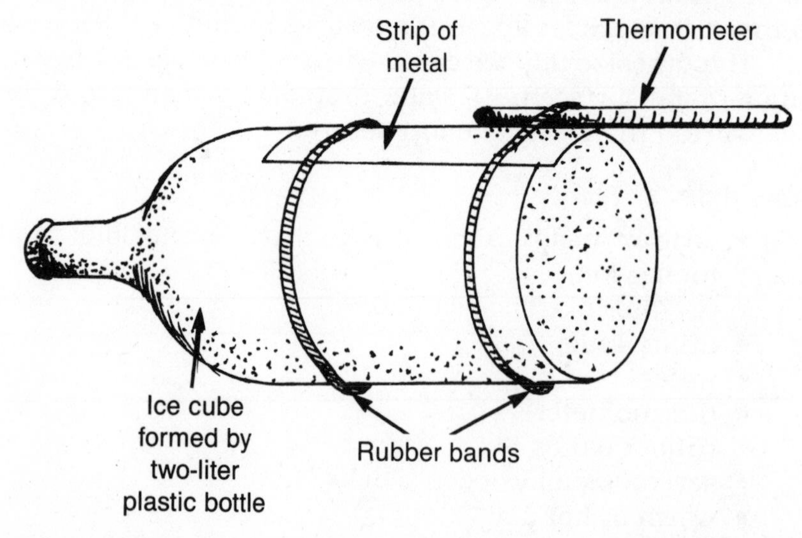

Fig. 8-3. *Make a large ice cube from a plastic two-liter soda bottle mold. Attach a strip of metal and a thermometer to observe dew point.*

might want to use several rubber bands and to place a thick piece of cloth or cotton over the top of the thermometer's bulb end. Air should be kept away from the bulb tip, as the temperature reading may be affected by exposure to the air.

Watch the metal strip as the ice cools it. When you see droplets of water appear on the metal strip, the surface of the strip has reached the dew point. Read the thermometer and record your results. Conclude whether your hypothesis was correct.

Going Further

1. Gather a plastic plate, thin sheet of wood, aluminum pie plate, paper plate, and metal tray of approximately equal thicknesses. Set them outside on an evening when you expect dew to occur. Very early the next morning, examine the objects. Is there dew on the ground? If so, which objects have dew on them? Can this experiment be quantified? When did the dew form (what time)? Construct a device to detect when moisture occurs (VCRs—video tape recorders—have "dew sensors" in them).
2. Try to perform the giant ice cube project on a cold day to see if you can determine the frost point.

PROJECT 8-3

How Wet Is the Air?

Adult Supervision Required

Overview

Moisture is in the air. Air temperature determines how much moisture the air can hold. Hot, muggy days occur when the air, as it warms, holds more water vapor. Cold days are seldom muggy. The air cannot hold much moisture. Relative humidity is the term used to indicate the percentage of moisture the air is holding compared to what it can hold. Guess what the relative humidity is in a room and then measure it. Hypothesize that you can increase it substantially.

Materials

- electric tea kettle (**adult supervision: boiling water**)
- thermometer
- string
- piece of medical gauze
- 1,000 milliliters of water
- small room
- rubber bands
- pad and pencil

Procedure

Fill an electric tea kettle with 1,000 milliliters of water, which is about as much as one can hold. Cut a small piece of medical gauze and soak it in water. Take these materials along with a pad and pencil into a small room, such as a den or bedroom, and shut all the doors and windows.

After being in the room for a few minutes, read and record the temperature of the air in the room. Then, using a rubber band, attach the piece of wet gauze to the bulb end of the thermometer. Tie a piece of string to the thermometer. Swing it around over your head 50 times. Evaporation will produce a cooling effect and the thermometer will have a lower reading on it than it did before. Record this temperature. Subtract the difference between the two readings. Use the chart shown in Fig. 8-4 to determine the relative humidity.

Relative Humidity Table

Dry Bulb	Difference between dry bulb and wet bulb (measured in degrees Fahrenheit) 1	2	3	4	5	6	7	8	9	10	11	12	13	14	15
64	95	90	84	79	74	70	65	60	56	51	47	43	38	34	30
66	95	90	85	80	75	71	66	61	57	53	48	44	40	36	32
68	95	90	85	80	76	71	67	62	58	54	50	46	42	38	34
70	95	90	86	81	77	72	68	64	59	55	51	48	44	40	36
72	95	91	86	82	77	73	69	65	61	57	53	49	45	42	38
74	95	91	86	82	78	74	69	65	61	58	54	50	47	43	39
76	96	91	87	82	78	74	70	66	62	59	55	51	48	44	41
78	96	91	87	83	79	75	71	67	63	60	56	53	49	46	43
80	96	91	87	83	79	75	72	68	64	61	57	54	50	47	44
82	96	92	88	84	80	76	72	69	65	61	58	55	51	48	45
84	96	92	88	84	80	76	73	69	66	62	59	56	52	49	46
86	96	92	88	84	81	77	73	70	66	63	60	57	53	50	47
88	96	92	88	85	81	77	74	70	67	64	61	57	54	51	48
90	96	92	89	85	81	78	74	71	68	65	61	58	55	52	49
92	96	92	89	85	82	78	75	72	68	65	62	59	56	53	50
94	96	93	89	85	82	79	75	72	69	66	63	60	57	54	51

Fig. 8-4. *Chart for calculating relative humidity.*

Have an adult boil off all of the water in the tea kettle, adding water vapor to the air in the room. The temperature in the room should still be the same. Wet the piece of gauze and again swing the thermometer with the gauze attached to it. Read the thermometer and use the chart to determine the new relative humidity of the room. Conclude whether your hypothesis was correct.

Going Further

1. Pick a cold room. If it's winter, turn off the heat to the room. If it's summer, run an air conditioner. Decrease a room's temperature and measure it. Measure the volume of the room, length times width, times height. Measure the relative humidity. Bring the temperature of the room up fifteen degrees. How many milliliters of water must you boil for that volume of room to bring the humidity up to what it was when the temperature was cooler? The air at the higher temperature can hold more than it can at the lower temperature. Air becomes dryer as temperature increases.

2. Determine the range of relative humidity at room temperature (70 degrees), where most people feel comfortable. This can be done by using several people and recording their responses. Using a tea kettle, increase moisture and measure the relative humidity. In the winter, homes are closed up tight, and heat from heating ducts and wood stoves makes the air drier. Consider comfort factors. For example, if a person's lips are chapped and they have been indoors all day, then the humidity level must be uncomfortably low.

STOP

PROJECT 4
The Pressure's On
Adult Supervision Required

Overview

Knowing if the barometric pressure is rising or falling can be important in predicting the weather. In this project, we will construct three homemade devices that detect a change in pressure. Examine the concepts of each and hypothesize which one will more accurately show a change in pressure. Which one will be the most reliable?

Materials

- two coffee cans
- masking tape
- balloon
- soda straw
- toothpick
- piece of oak tag or stiff cardboard
- 18-inch-long glass tube with one end sealed
- two-liter plastic soda bottle
- one-hole stopper to fit in the mouth of a two-liter soda bottle
- piece of clear, flexible tubing
- short piece of glass tube (to go through the stopper and connect with the flexible tubing—the glass part of an eyedropper will work)
- rubber bands
- string
- water
- food coloring
- glue
- cork
- something to punch several holes in the coffee can (adult supervision required)

Procedure

First, construct the membranelike device shown in Fig. 8-5. This is done by stretching a piece of balloon, or rubber dam, which is available from scientific supply firms listed in

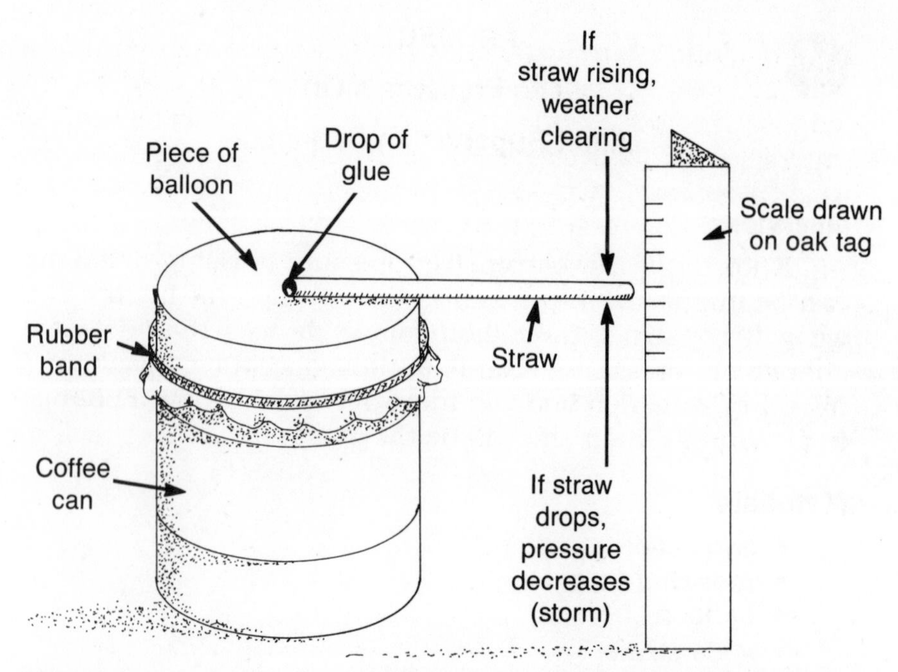

Fig. 8-5. *A homemade barometer (with a rubber membrane top) monitoring trapped air inside a can.*

the Resource List, over a coffee can. The coffee can's bottom must be tightly sealed. The rubber piece on top must seal the air inside. Glue can be placed all around the top of the can. The rubber should hang about one inch over the sides. Use string or rubber bands to secure the rubber membrane. It should remind you of an Indian tom-tom.

Take a plastic soda straw and glue a toothpick in one end. This will act as an accurate pointer. Instead of a plastic soda straw, you could use a seven- or eight-inch-long strand of straw from a kitchen broom.

Put one end of the straw in the middle of the rubber membrane and place a drop of glue on the end to attach it. Lay the straw flat so that it hangs off the end of the coffee can by several inches. Draw a scale on a piece of oak tag or sturdy cardboard. The scale should have lines at 1/8 of an inch intervals. The middle of the scale should be marked 0 (zero) and each line below it should be marked −1, −2, −3, and above it 1, 2, 3, and so on. Fold the paper so it will stand up or somehow support it so the straw acts like an indicator, pointing to the scale.

When pressure increases, the middle of the membrane will go down, causing the indicator that is resting on the edge of the coffee can to rise. When pressure decreases, the middle of the membrane will go up, causing the indicator to point downward. Obviously, the can should be sealed on a day when the pressure is not at extremes. Do not seal it on a stormy day, for instance. Otherwise, the indicator will not have as much range.

Next, construct the column of water device shown in Fig. 8-6. Fill a long glass tube, about 18 inches long, with colored water. The tube should be closed at one end. Put a cork in

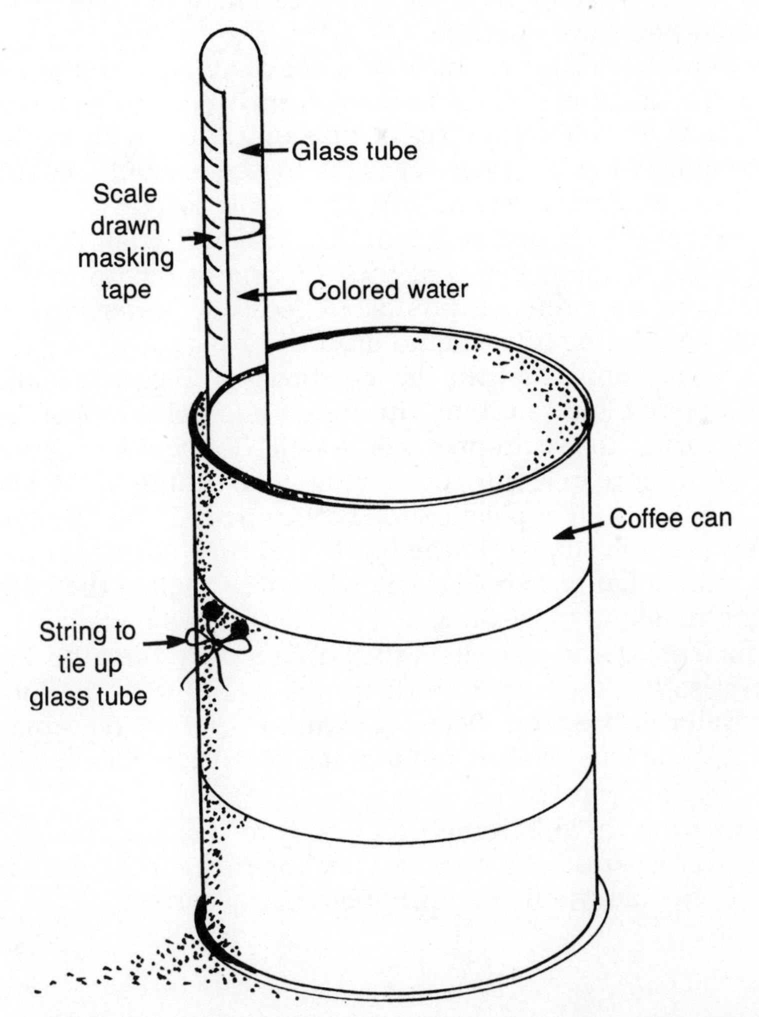

Fig. 8-6. *A homemade barometer using an upside-down glass indicator tube.*

the mouth of the tube to keep the water from coming out. Punch two small holes in the coffee can, about two inches down from the top. The column of water will be placed upside down in the coffee can and will need to be tied up and secured to the wall of the can. Fill the coffee can three-quarters full of colored water. Turn the water-filled tube upside down so that the cork is under the water in the can. Take the cork out. The water should remain in the tube. Secure the tube against the side of the coffee can by running string or rubber bands through the holes punched in the side. It is important to keep the open mouth of the tube suspended a little above the bottom of the coffee can so water will be free to enter and leave the tube.

Using a soda straw, blow bubbles of air up into the tube until the water level drops in the column about three inches.

Draw a scale on a strip of masking tape, with lines at one-eighth-inch intervals. Paste it to the side of the tube, with the water level in line with the zero mark on the scale.

When the air pressure is high, it should push down on the water in the coffee can, thus forcing water up into the tube. Low air pressure would release tube water, and the water level in the tube should drop.

Finally, construct the device shown in Fig. 8-7. Place a small piece of glass tubing through a one-hole stopper. The glass part of an eye dropper works well. Connect a long piece (a foot or more in length) of clear, flexible tubing to the glass tube. Fill a two-liter plastic soda bottle one-half full of colored water. Put the stopper in the bottle and turn it upside down. Use rubber bands to hold the flexible tube against the bottle. Using masking tape, draw a scale, with lines one-eighth of an inch apart and place it on the bottle next to the tube. Position it so the zero mark on the scale is level with the top of the water in the tube. Devise a stand to support the upside-down apparatus. Be sure not to put a kink in the flexible tubing.

Use the three barometric devices to measure the pressure over a period of a week or two. Log measurements each day. Conclude whether your hypothesis is correct.

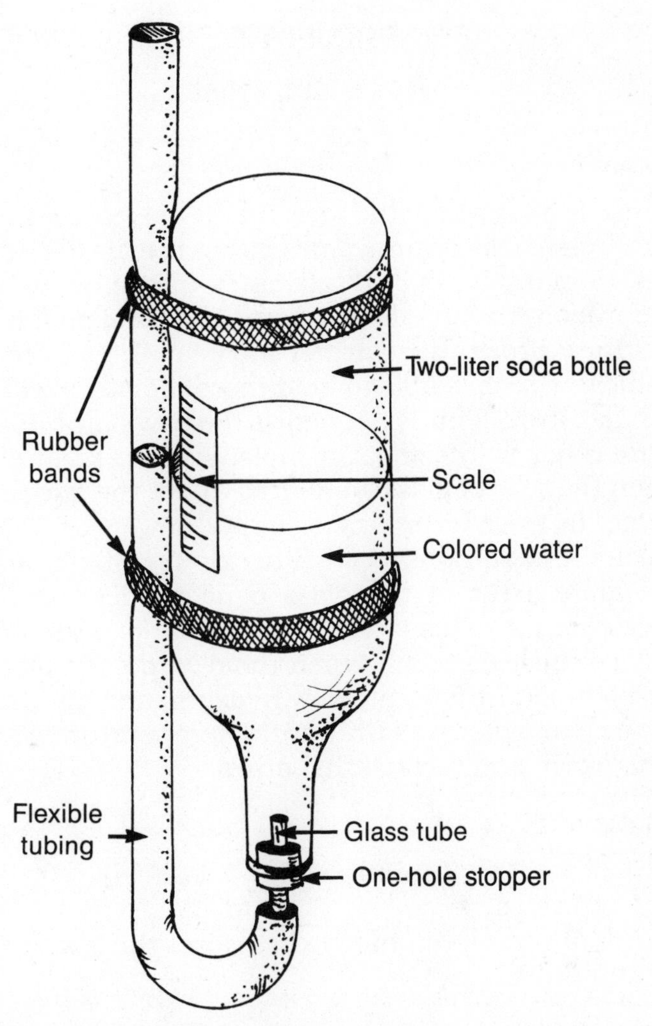

Fig. 8-7. *A homemade barometer using an upside-down plastic soda bottle and a piece of flexible tubing.*

Going Further

1. Compare the measurements made by the home-made barometers to a commercially available one. Which ones change sooner? Which show the greatest change?
2. As pressure increases, clouds usually disappear. Is there a relationship between pressure and humidity?

PROJECT 5
Jack's Yard Frost

Overview

Frost forms when the dew point is reached below 0 degrees Celsius. Moisture comes out of the air in the form of a solid. In many parts of the country during winter, frozen layers form on the outside of automobile windshields. As the temperature drops, the warmer air near the surface of the windshield, which is able to hold more moisture, condenses to form moisture. This is defined as the dew point. If the temperature is below freezing, the moisture forms as frost.

Using a piece of glass supported above the surface of the ground, will frost form on the upper, lower, or both sides (given all the conditions for frost to exist)? Will the frost form in the same areas of the glass pane as dew would on a warmer evening? This would require one experiment to be conducted during a night when the dew point was reached and the temperature went below freezing, and another night when the dew point was hit but the temperature was above freezing. Form and record a hypothesis.

Materials

- pane of glass (minimum size 1 x 2 feet)
- four bricks
- minimum thermometer that stores the lowest temperature until reset

Procedure

Using four bricks as supports, suspend a pane of glass above the ground surface (see Fig. 8-8). If the glass has sharp edges, have an adult wrap tape around them or enclose the sharp edges with some protecting cover.

Place the setup outside when evening approaches. Early in the morning, before the sun has had a chance to evaporate any dew, examine the ground and surrounding objects. If they have dew on them, then the dew point was reached and you can examine the glass plate for condensation. You can assume that the temperature was above the freezing point. Repeat this procedure for a night when the temperature goes

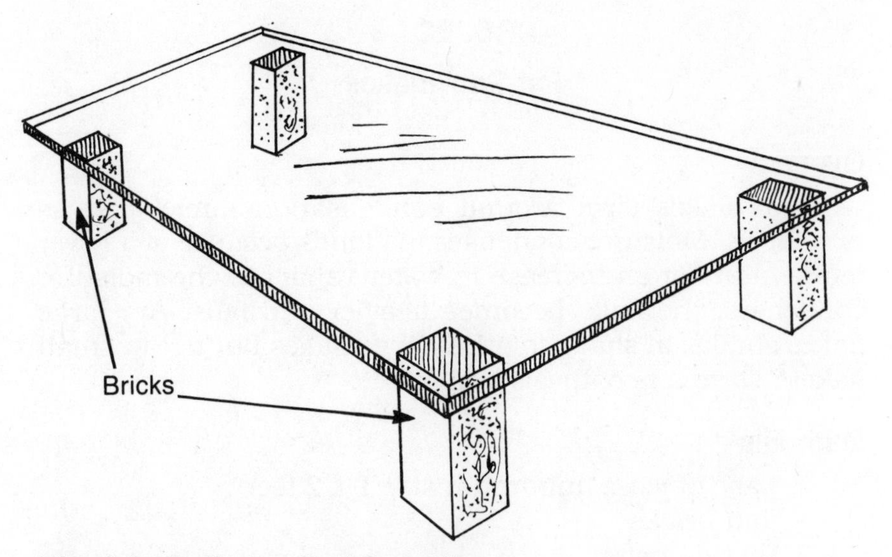

Fig. 8-8. *Exploring frost formation.*

below freezing. Examine for frost. Study your data and conclude whether your hypothesis was correct.

Going Further

1. Do any conditions occur where formations happen on both sides, and at other times only form on one side?
2. Is there a difference in location (concrete, grass, patio, deck, driveway, etc.)? Use two sets to make comparisons on the same nights.
3. Can the distance from the ground cause a change? What if the glass is placed vertically instead of horizontally? Will tape or plastic wrap affect formations?

PROJECT 6
Pet Snowflakes

Overview

Snowflakes form around condensation nuclei just as rain forms. Moisture condenses in clouds because of a lower temperature or an increase in water vapor. As the moisture condenses, the flake becomes heavier and falls. Are large flakes similar in shape to other large flakes but not to small flakes? Form a hypothesis and test it.

Materials

- pane of glass (minimum size 1 x 2 feet)
- four bricks
- thermometer
- magnifying glass
- black construction paper

Procedure

Use a cooled pane of glass and a cooled magnifying glass to catch and evaluate snowflakes. Suspend the pane above the ground as shown in Fig. 8-9. Draw sketches. Measure the temperature. Record all data. During another snowfall, again catch and examine snowflakes. Compare them to the data you previously obtained.

Going Further

1. Does shape or size relate to temperature or humidity? Keep data on shape, size, and temperature. Can you separate individual snowflakes?
2. What conditions exist in your area for snow? What conditions produce hail or sleet?
3. Can fog be present below zero degrees Celsius?

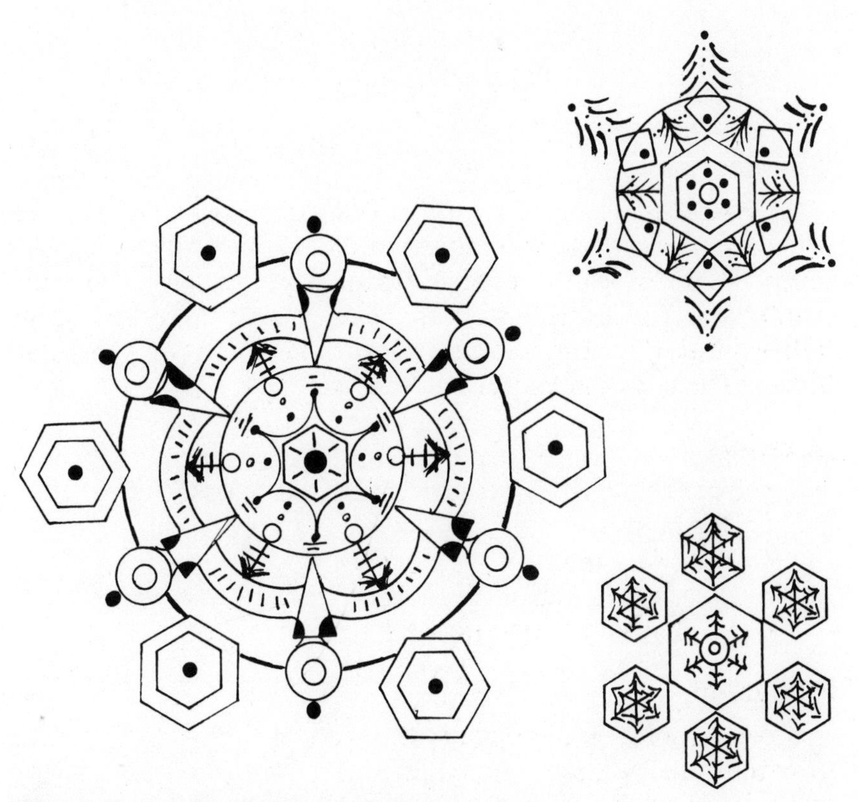

Fig. 8-9. *Evaluating snowflake patterns.*

Glossary

accretion—The slow, steady buildup of materials.
archaeology—The scientific study of objects from the past.

climate—The long-term weather average of a large geographic area.
control group—When doing experiments, a control group is the group that has all the variables maintained. For example, if you want to test for the effects of carbon monoxide on plants, you must have two equally healthy plants. Both plants will receive exactly the same care and conditions (soil, sunlight, water) and one plant, the experimental plant, will receive additional carbon monoxide. The other plant is the control plant. The control plant is maintained while the experimental plant receives the variation.
cryogenics—The study of the effects of low temperatures on objects and processes.
crystals—Minerals whose atoms are arranged in a repeating pattern. Examples of crystal structures are sugar, copper sulfate, and quartz.

dew point—The temperature at which moist air must be cooled for saturation to occur. If the temperature is below freezing, the dew point is sometimes called the frost point.

erosion—The wearing away of a material. Erosion can be caused by friction, water, ice, wind, sand, chemicals, and temperature extremes.

experiment—A planned way to test a hypothesis.

fossils—Are evidence of things once alive, and can be either the preserved organism itself, an imprint of the organism (leaf, footprint), or a cast where the organic material has been replaced by minerals (petrified wood).

Fresnel lens—A lens that concentrates light.

frost line—The depth at which frost is virtually nonexistent.

hypothesis—A theory or educated guess. "I think when asked how much they would weigh on Mars, more boys will have accurate guesses than girls."

igneous rocks—Rocks formed by a cooling process, such as hot lava from a volcano.

infrared—Light that cannot be seen with the naked eye and is below the color red outside the visible spectrum of light.

lithosphere—The earth's crust, which is about 25 miles thick and 8,000 miles in diameter.

metamorphic rocks—Rocks that were once either igneous or sedimentary rocks that have undergone change due to heat.

minerals—Naturally occurring, inorganic material (they are not alive nor do they come from living things), made up of one or more elements (such as silicon, oxygen, iron). Combinations of minerals make up rocks.

observation—Looking carefully.

Pangaea—The seven continents on the earth might have once been joined to make up one huge land mass scientists call Pangaea.

passive solar—Devices that collect energy from the sun and release heat without requiring any additional energy to work.

photoluminescence—Luminescence caused by the absorption (takes in) of infrared radiation, visible light, or ultraviolet light.

photovoltaic cells—Semiconductor devices that change light from the sun directly into electricity.

plate tectonics—Sections or "plates" of the earth's crust that are drifting on the liquid core of the earth.

quantify—To measure.

quartz—The most abundant mineral on earth (beach sand).

relative humidity—The percentage of moisture the air is holding.

rocks—Rocks are combinations of one or more minerals, and make up the solid part of earth.

sample size—The number of items under test. The larger the sample size, the more significant the results. Using only two plants to test a hypothesis that sugar added to water results in better growth would not yield a lot of confidence in the results. One plant may grow better simply because some plants just grow better than others.

sandstone—An accumulated material formed by layering and compacting.

scientific method—A step-by-step logical process for investigation. A problem is stated, a hypothesis is formed, an experiment is set up, data is gathered, and a conclusion is reached about the hypothesis based on the data gathered.

sedimentary rocks—Rocks formed by a layering of material that settles.

seismic waves—Vibrations traveling through the earth.

snow fences—A fence designed to protect buildings, roads or railroad tracks from winds carrying drifting snow by disrupting the wind flow and causing it to deposit the snow on the leeside of the fence. Windbreaker fences are often used near seashores to prevent beach erosion.

solar energy—Heat generated by the sun.

specific gravity—The comparison between an unknown and equal volume of water by weight.

supersaturated—A point at which a solution can no longer hold any additional material.

tensile strength—The greatest lengthwise strength an object can bear without tearing apart.

thermal energy—Heat produced from inside the earth.

ultraviolet—Light that cannot be seen with the naked eye and is above the color violet in the visible light spectrum.

Venturi effect—As moving air is squeezed through a small opening, its velocity increases.

visible light spectrum—Seven colors that are visible to the human eye and include red, orange, yellow, green, blue, indigo, and violet.

weather—The daily changes in the local atmosphere, caused by moving air and gases, tiny particles, and water vapor (wind, rain, snow, fog).

Resource List

This resource list is compiled to give you a mail-order source for science supplies. Each address has been checked for accuracy.

Carolina Biological Supply Company
2700 York Road
Burlington, North Carolina 27215
1-800-547-1733

Edmund Scientific Company
101 E. Gloucester Pike
Barrington, NJ 08007
609-573-6250
Free catalog available

Fisher Scientific
4901 W. LeMoyne St.
Chicago, IL 60651
1-800-621-4769

Frey Scientific Company
905 Hickory Lane
Mansfield, OH 44905
1-800-225-FREY

Science Kit & Boreal Laboratories
777 East Park Drive
Tonawanda, NY 14150-6782
1-800-828-7777

Sargent-Welch Scientific Company
7300 North Linder Ave.
PO Box 1026
Skokie, IL 60077
312-677-0600

Heath Company
Benton Harbor, Michigan 49022

Sells electronic equipment, weather instruments, computers, test equipment

Index

F

G

H

I

J

L

M

O

P

About the Authors

Bob Bonnet holds an MA degree in environmental education, and has been teaching science at the junior high school level in Dennisville, New Jersey for more than fifteen years. He is also a State Naturalist at Belleplain State Forest in New Jersey. During the last seven years, he has organized and judged many science fairs at the local and regional levels. Mr. Bonnet is currently the chairman of the Science Curriculum Committee for the Dennisville School system.

Dan Keen holds an Associate in Science Degree, majoring in electronic technology. Mr. Keen is a computer consultant who has written many articles for computer magazines and trade journals since 1979. He is the coauthor of two computer books published by TAB BOOKS, Inc., Mastering the Tandy 2000 and Assembly Language Programming for the TRS-80 Model 16. *In 1986 and 1987 he taught computer science at Stockton State College in New Jersey. His consulting work includes writing software for small businesses and teaching adult education classes on computers at several schools.*

Together, Bob Bonnet and Dan Keen have published articles on a variety of science topics.